MW00657721

F
A
79
1/

Praise for
Evolution: A View from the 21st Century

"A novel description of evolution based on modern genetic analysis is presented in detail. Shapiro has written a stimulating, innovative manuscript that surely Darwin would have liked."

—**Sidney Altman**, Yale University;
Nobel Laureate in Chemistry, 1989

"Based on a long and highly competent personal experience in science and his novel insights into biological functions, the author has reached views of biological evolution that can reveal to a wide, interested readership how the living world co-evolves with the environment through its intrinsic powers."

—**Werner Arber**, Professor Emeritus, University of Basel, Switzerland;
Nobel Laureate in Physiology/Medicine, 1978

"Professor Shapiro's offering is the best book on basic modern biology I have ever seen. As far as I can tell, the book is a game changer."

—**Carl Woese**, University of Illinois; discoverer of Archaea, the third realm of life;
National Medal of Science 2000

"This moving, clearly written, if complex, description of 'natural genetic engineering' explains evolutionary processes that preceded people by at least 3,000 million years. Shapiro's detailed account of ubiquitous genetic dynamism, DNA machination, repair, and recombination in real life, bacterial to mammalian, destroys myths. Genes, whatever they are, are not 'eternal.'... Shapiro's careful, authoritative narrative...is entirely scientific and should interest all of us who care about the evolution of genetic systems."

—**Lynn Margulis**, University of Massachusetts, Amherst;
National Academy of Sciences, National Medal of Science 1999

"From revisiting the Central Dogma, to outlining a systems biological approach to evolution, this book is a magnificent analysis of the key questions of the origin of variation and, therefore, of evolutionary change. Starting from his early encounter with the work and ideas of Barbara McClintock, through to his work on systems engineering as the key to understanding organisms, Jim Shapiro has new insights on all the central issues of evolutionary theory. The genome becomes a read-write storage system rather than the sole determinant of heredity. After reading this book, you will find it imperative to see biology as the 21st century is coming to see it. The ambitious title of the book is fully justified.

This book gives solid biological evidence on the origin of variation and evolutionary change. That evidence will surprise those who still think that variation is random. Cells and organisms sense their environment and transmit that information to their genomes. Brilliant."

—**Denis Noble**, CBE FRS, Balliol College, Oxford;
author of *The Music of Life*

"Evolution, or constant change in nature, is the deepest of human ideas. It is the core of all physical, biological, and social systems and sciences—linking energy, matter, life, and consciousness. The dramatic genomic revolution unfolded new informative horizons of constant combinatorial diversity, change, and emerging innovation from simple to complex in biological evolution. Much of the understanding of genome complex adaptive architecture, dynamic reorganization, expression, and regulation—as well as the origin of life, cells, and biochemical networks—is a future challenge. This book highlights the exciting challenges to explore future nonconventional and nondogmatic vistas of evolutionary biology. The explorative mysteries include roles in evolution of nonrandom adaptive mutations, epigenetics, and repetitive DNA function and regulation. All contribute to dynamic systems biology and engineering between the evolving genome, cell, and environmental stresses and changes affecting the dynamic genome read-write memory system underlying life's evolution."

—**Eviatar Nevo**, University of Haifa, U.S. National Academy of Sciences; explorer of Evolution Canyon

Evolution

A View from the 21st Century

James A. Shapiro

Vice President, Publisher: Tim Moore
Associate Publisher and Director of Marketing: Amy Neidlinger
Acquisitions Editor: Kirk Jensen
Editorial Assistant: Pamela Boland
Development Editor: Russ Hall
Senior Marketing Manager: Julie Phifer
Assistant Marketing Manager: Megan Colvin
Cover Designer: Chuti Prasertsith
Managing Editor: Kristy Hart
Project Editor: Anne Goebel
Copy Editor: Gayle Johnson
Proofreader: Linda Seifert
Indexer: Erika Millen
Senior Compositor: Gloria Schurick
Manufacturing Buyer: Dan Uhrig

FT Press Science offers excellent discounts on this book when ordered in quantity for bulk
purchases or special sales. For more information, please contact U.S. Corporate and
Government Sales, 1-800-382-3419, corpsales@pearsontechgroup.com. For sales outside the
U.S., please contact International Sales at international@pearson.com.

First Printing June 2011

ISBN-10: 0-13-278093-3
ISBN-13: 978-0-13-278093-3

Pearson Education LTD.
Pearson Education Australia PTY, Limited
Pearson Education Singapore, Pte. Ltd.
Pearson Education Asia, Ltd.
Pearson Education Canada, Ltd.
Pearson Educación de Mexico, S.A. de C.V.
Pearson Education—Japan
Pearson Education Malaysia, Pte. Ltd.

Library of Congress Cataloging-in-Publication Data:

Shapiro, James Alan, 1943-
 Evolution : a view from the 21st century / James A. Shapiro.
  p. cm.
 ISBN 978-0-13-278093-3 (hardback : alk. paper)
 1. Evolution (Biology) 2. Genomics. 3. Molecular genetics. 4. Evolutionary genetics. I.
Title.
 QH447.S53 2012
 576.8—dc22
                              2011008338

This book is dedicated to Felix and Gus and
all future grandchildren.
May their generation share nature's deep wisdom in
adapting life to its home on Earth.

Contents

Acknowledgments xiii

About the Author xv

A Note on Reading This Book for Individuals
with Different Backgrounds xvii

Introduction: Taking a Fresh Look at the Basics
of Evolution in the New Century 1

**Part I   Sensing, Signaling, and Decision-Making in
Cell Reproduction 7**

How *E. Coli* Chooses the Best Sugar to Eat 8

Proofreading DNA Replication 12

DNA Damage Repair and Mutagenesis 14

Cell Cycle Checkpoints .18

Signaling from the Cell Surface to the Genome:
Pheromone Response in the Sexually Aroused
Yeast Cell 21

The Role of Intercellular Signals in the Cell Death
Decision 23

Revisiting the Central Dogma of Molecular Biology . . .24

**Part II   The Genome as a Read-Write (RW) Storage
System 27**

Genome Formatting for Proper Access to Stored
Information 30

Genome Compaction, Chromatin Formatting,
and Epigenetic Regulation 31

Genome Formatting for Replication, Localization,
and Transmission to Daughter Cells 36

Distinct Classes of DNA in the Genome 39

The Molecular Mechanisms of Natural Genetic
Engineering 43

Natural Genetic Engineering as Part of the Normal
Life Cycle 55

Cellular Regulation of Natural Genetic Engineering . . .69

Targeting of Natural Genetic Engineering within the
Genome 82

Reviewing What Cells Can Do to Rewrite Their
Genomes over Time 87

Part III   **Evolutionary Lessons from Molecular Genetics**
                 and Genome Sequencing 89
                 Antibiotic Resistance and Horizontal DNA Transfer . . .90
                 The Modular and Duplicative Nature of
                 Protein Evolution 95
                 Molecular Taxonomy and the Discovery of a
                 New Cell Type 98
                 Symbiogenesis and the Origin of Eukaryotic Cells . . .100
                 Natural Genetic Engineering and Evolutionary
                 Genomic Innovation 107
                 Use and Reuse of Evolutionary Inventions 115
                 What Makes a Man Different from a Mouse? 118
                 Whole Genome Doubling at Critical Stages of
                 Evolutionary Innovation and Divergence 120
                 Reviewing What the DNA Record Reveals about
                 Cell Activities over Evolutionary Time 124

Part IV   **A New Conceptual Basis for Evolutionary**
                 Research in the 21st Century 127
                 A Systems Approach to Generating Functional
                 Novelties 129
                 Reorganizing Established Functions to Generate
                 Novelty 130
                 Generation of Novel Components 131
                 Retention, Duplication, and Diversification of
                 Evolutionary Inventions 133
                 The Implications of Targeting Genome
                 Restructuring 134
                 Can Genomic Changes Be Linked to Ecological
                 Disruptions? 139
                 What Might a 21st Century Theory of Evolution
                 Look Like? 142
                 Where Does Evolution Fit in 21st Century Science? . . .145

                 Glossary **149**

                 References **175**

                 Index **241**

The following material can be found online only (www.ftpress.com/shapiro):

Appendixes to Part I and Part II of the Printed Book

Appendix I.1. The *lac* operon control circuit.

Appendix I.2. *S. cerevisiae* pheromone response signal network (Sprague 1991).

Appendix II.1. Formatting the genome for transcriptional control.

Appendix II.2. Natural genetic engineering in B lymphocytes for the rapid evolution and maturation of a virtually infinite diversity of antigenbinding proteins.

References

Table References for Part II and Part III of the Printed Book

Table II.1 DNA content in higher eukaryotes (Shapiro and Sternberg 2005)

Table II.2 Different classes of annotated repetitive genome components (amplified from (Shapiro and Sternberg 2005))

Table II.3 Generic Cell Operations That Facilitate DNA Restructuring [30, 61, 392] *Note: These references are listed in the printed book only.*

Table II.4 Natural Genetic Engineering Systems [6-8] *Note: These references are listed in the printed book only.*

Table II.5 Control of bacterial protein synthesis (phase variation) and modification of protein structure (antigenic variation) by natural genetic engineering (expanded from (Wisniewski-Dye and Vial 2008))

Table II.6 Applications of site-specific recombination to different functions in bacterial cells (Hallet and Sherratt 1997)

Table II.7 Various stimuli documented to activate natural genetic engineering

Table II.8 Genomic responses to changes in ploidy and interspecific hybridization in plants and animals

Table II.9 RNA-based defense against viruses and plasmids in bacteria, archaea, fungi and plants

Table II.10 Life history events that alter the epigenome (DNA methylation and chromatin formatting)

Table II.11 Examples of targeted natural genetic engineering

Table III.1 Examples of intercellular and interkingdom DNA transfer

Table III.2 Natural genetic engineering documented in the evolution of sequenced genomes

Suggested Readings for Non-Professionals
Introduction Readings
The traditional perspective

Heredity outside the genome

New conceptual approaches
Part I Readings
How *E. coli* chooses the best sugar to eat

Proofreading DNA replication

DNA damage repair and mutagenesis

Cell cycle checkpoints

From the cell surface to the genome

The role of intercellular signals in the cell death decision

Revisiting the central dogma of molecular biology
Part II Readings
Genome formatting for properly accessing stored information

Genome compaction, chromatin formatting, and epigenetic regulation

Genome formatting for replication and transmission to daughter cells

Distinct classes of DNA in the genome

The molecular mechanisms of natural genetic engineering

Natural genetic engineering as part of the normal life-cycle

Developmental, cell, and molecular biology of the adaptive immune system

Cellular regulation of natural genetic engineering
Part III Readings
Antibiotic resistance and horizontal DNA transfer

The modular and duplicative nature of protein evolution

Nucleic acid sequencing, molecular taxonomy, and the discovery of a new cell type

Cell evolution: mutalism, parasitism, pathogenesis, endosymbiosis, bacterial multicellularity, symbiogenesis, and the origin of eukaryotic cells

Natural genetic engineering and genomic innovation

Use and reuse of evolutionary inventions:
Basic biology of morphogenesis and multicellular development

Molecular biology of animal and plant development

Evolutionary combinatorics and reuse

Whole genome doubling at critical stages of evolutionary innovation and divergence

Part IV Readings

Reorganizing established functions to generate novelty

The implications of targeting genome restructuring

Can genomic changes be linked to ecological disruptions?

Biogeochemical influences of evolution on the environment

Paleontological record and mass extinctions

What might a 21st Century theory of evolution look like?

Where does evolution fit in 21st Century science?

Molecular biology background readings in chronological order

References for the Printed Book

Extra References

Cell Cycle Checkpoints

Cellular Regulation of Natural Genetic Engineering

DNA Damage Repair and Mutagenesis

From the Cell Surface to the Genome

The Genome as a Read-Write (RW) Storage System

How *E. Coli* Chooses the Best Sugar to Eat

Molecular Mechanisms of Natural Genetic Engineering

Proofreading DNA Replication

Symbiogenesis and the Origin of Eukaryotic Cells

What Makes a Man Different from a Mouse

Acknowledgments

My deepest debt of gratitude belongs to Barbara McClintock for sharing her friendship and wise scientific insights with me over the last dozen-plus years of her life. She opened my eyes to a freer way of thinking about science in general and evolution in particular. Next comes Bill Hayes, my doctoral supervisor, who taught me the virtues of experimental simplicity, close observation, and logical thinking without preconceptions. He was an outstanding example of humane mentoring in an otherwise highly competitive field of scholarship.

This book had its genesis in two critical events separated by 15 years. In 1993, I was lucky enough to win the Darwin Prize Visiting Professorship from the University of Edinburgh. The accompanying lectures induced me to begin thinking in a comprehensive way about the role of natural genetic engineering in the evolutionary process. I owe the invitation chiefly to two couples, my longtime friends Willy and Millie Donachie and my newer friends Ken and Noreen Murray. The second event came in 2008 when a German colleague, Joachim Bauer, invited me to collaborate on an English version of his book, *Das Kooperative Gen*. Although that collaboration did not prove feasible, Joachim's invitation stimulated me to become serious about completing a book on new ways of thinking about evolution. The 15-year wait was well worth it. An enormous amount of useful molecular genetic and genomic data became available in the interval. These data documented and reinforced views I had developed in my conversations with McClintock and others.

Among the other colleagues who were not direct collaborators but who helped form my views about conceptual developments in biology and details of genome changes, I have to single out Asad Ahmed, Guenther Albrecht-Buehler, Michael Ashburner, Dennis Bray, Sydney Brenner, Gerard Buttin, Allan Campbell, Nancy Craig, Naomi Datta, Harvey Eisen, Hannah Engelberg-Kulka, Shelly Esposito, Michel Faelen, Alex Frisch, Misha Golubovsky, Dan Gottschling, Jim Haber, Fred Heffron, Maurice Hofnung, John Holland, Lynn Margulis, Alfonso Martinez-Arias, Max Mergeay, Matt Meselson, Kiyoshi Mizuuchi, Jacques Monod, Dennis Noble, Phoebe Rice, Ethan Signer, Hewson Swift, Ed Trifonov, Adam Wilkins, and members of the 1979 Nicholas Cozzarelli lab at the University of Chicago, who rigorously but gently critiqued early versions of my molecular model for phage Mu transposition. I have to beg the indulgence of many other colleagues whose interactions and discoveries helped shape my views and hope they will forgive me for not mentioning their names; doing so would make these acknowledgments look like the speaker list of a large international conference.

To my direct collaborators over more than 40 years of research on genome engineering in *E. coli*, I owe special thanks to Sankhar Adhya, Jon Beckwith, Mike Benedik, Spencer Benson, Gary Boch, Pamela Brinkley, Ahmed Bukhari, Carol Burck, Malcolm Casadaban, Stan Cohen, Nancy Cole, Marty Dworkin, Michel Faelen, Mike Fennewald, Bernhard Hauer, Pat Higgins, Garret Ihler, Karen Ippen-Ihler, Viktor Krylov, Roxanne Laux, David Leach, Lorne MacHattie, Genevieve Maenhaut-Michel, Dani McBeth, Richard Meyer, Carole-Jean Muster, David Owen, Shenaz Rehmat, Peter Sporn, Arianne Toussaint, Rick von Sternberg, and Bernard Witholt. Together with them, I found that *E. coli* and other organisms held many surprises about what they could do to and with their genomes. Genevieve deserves special distinction for the impressive and highly original series of experiments she performed in the 1990s that clarified many aspects of how *E. coli* cells responded to oxidative starvation.

I wish to express my gratitude to the organizations that supported my studies and research in bacterial genetics and transposable elements, as well as my thinking on evolutionary processes: The Marshall Aid Commemoration Commission of the United Kingdom, the Wellcome Trust, The Jane Coffin Childs Memorial Fund, the American Cancer Society, the National Science Foundation, the National Institutes of Health, the American Chemical Society, the Louis Block Fund of the University of Chicago, and the Darwin Prize Trust of the University of Edinburgh.

In the preparation of this book, the help and advice of my literary agent, Georges Borchardt, and of my editor, Kirk Jensen, proved invaluable. Without their help and guidance, this volume would not have seen the light of day. As always, I am indebted to my wife, Joan, for allowing me the freedom to write this book in the summer of 2010, for her outstanding editorial skills, and for providing a loving and intellectually stimulating domestic environment. Finally, I have to thank my children, Jacob and Danielle, for taking such fine care of our home as teenagers while Joan and I were enjoying the 1993 Darwin Professorship in Edinburgh.

About the Author

James A. Shapiro is Professor of Microbiology at the University of Chicago. He has a BA in English Literature from Harvard (1964) and a PhD in Genetics from Cambridge (1968). William Hayes was his PhD supervisor, and Sydney Brenner was an unofficial adviser during his time in Cambridge as a Marshall Scholar. His thesis, *The Structure of the Galactose Operon in Escherichia coli K12*, contains the first suggestion of transposable elements in bacteria. He confirmed this hypothesis in 1968 during his postdoctoral tenure as a Jane Coffin Childs fellow in the laboratory of François Jacob at the Institut Pasteur in Paris. The following year, as an American Cancer Society fellow in Jonathan Beckwith's laboratory at Harvard Medical School, he and his colleagues used *in vivo* genetic manipulations to clone and purify the *lac* operon of *E. coli*, an accomplishment that received international attention. In 1979, he formulated the first precise molecular model for transposition and replication of phage Mu and other transposons. In 1984, he published the first example of what is now called "adaptive mutation." He found that selection stress triggers a tremendous increase in the frequency of Mu-mediated fusions. Together with Pat Higgins in 1989, he showed that activation of Mu replication and transposition is spatially organized in bacterial colonies. Since 1992, he has been writing about the importance of biologically regulated natural genetic engineering as a fundamental new concept in evolution science.

Following a teaching stint at the University of Havana (1970–1972) and research at Brandeis (1972–1973), Shapiro moved to a faculty position at the University of Chicago in 1973. He has been there since then with occasional sabbaticals and visiting professor appointments at the Institut Pasteur, Tel Aviv University, Cambridge University, and the University of Edinburgh, where he was the Darwin Prize Visiting Professor in 1993. He is a fellow of the American Academy of Microbiology, the AAAS and the Linnean Society of London. In 2001, he received an honorary O.B.E. from Queen Elizabeth for services to higher education in the UK and US.

Together with Ahmed Bukhari and Sankhar Adhya, Shapiro organized the first conference on DNA insertion elements in May 1976 at

Cold Spring Harbor laboratory. He is editor of *DNA Insertion Elements, Episomes and Plasmids* (1977 with Bukhari and Adhya), *Mobile Genetic Elements* (1983), and *Bacteria as Multicellular Organisms* (1997 with Martin Dworkin). From 1980 until her death in 1992, he maintained a close scientific and personal friendship with Barbara McClintock, whom he credits with opening his eyes to new ways of thinking about science in general and evolution in particular.

A Note on Reading This Book for Individuals with Different Backgrounds

I wrote this book to inform both professional biologists and interested nonprofessionals of the extensive series of developments in molecular genomics that require a deep rethinking of basic evolutionary concepts. By adopting a perspective that is, in many ways, counter to conventional wisdom, I accepted the obligation to document thoroughly the science behind my views. For the professional reader, this is the expected norm. However, for the nonprofessional reader, fulfilling this obligation means that the book goes into a great deal of technical detail that may sometimes seem opaque and confusing. To lessen the burden of coping with so many specific biological examples, I have tried to segregate much of the purely technical information, especially the references, into tables and appendixes, which are available at the book's website (www.ftpress.com/shapiro) and at my personal website (http://shapiro.bsd.uchicago). The references are far more useful online than in print because they are linked to abstracts and the articles themselves.

Despite the extensive online postings, this book still contains many discussions that may prove challenging for the general reader. The most difficult will be in Part II, which describes the genome as a Read-Write (RW) memory system and details the many ways cells inscribe information into their genomes at different time scales. This section is designed in part to pull together a small part of the thousands of studies which can reasonably be summarized as showing that these diverse inscriptions constitute a form of writing that modifies the cell's genomic memory. To my knowledge, there is no comparable review of this material. Thus, while general readers may find some of the detail excessive, my hope is that professional readers will view it as a valuable source of references that clarifies just how strong the evidence is for viewing the genome as a RW system.

For those hardy nonprofessional readers determined to complete the entire book, I suggest using two excellent online aids. You can use Wikipedia to look up terms and subjects that seem to pop up out of the ether (such as mitosis). Although variable, I have found Wikipedia an excellent source of explanation on many technical subjects. Additionally, in the book section of the PubMed biomedical database (http://www. ncbi.nlm.nih.gov/books), you can access a number of standard textbooks using a subject-based search. The textbook examples are well thought out and clearly illustrated. Both of these online resources provide the basis for a solid and rapid self-directed education in modern cell and molecular biology. Wikipedia is also an excellent source for exploring the historical development of evolutionary thought and the life sciences in general.

An additional autodidactic resource is a list of suggested readings intended for general audiences, found online on the author's and book's websites. Most of these suggested readings come from archived copies of *Scientific American*, available online (to subscribers or through subscribing institutions). The reading list also includes some newspaper articles and links to websites where the works of Darwin and Stephen J. Gould can be accessed. The suggested readings are organized according to the book's contents, and citations to them in the printed book appear in bold type to help provide well-articulated reference material suitable for the nonprofessional reader.

If you just want to get the gist of the argument without simultaneously taking a self-directed course in modern biology, I suggest reading the Introduction plus the beginning and concluding sections of each of the four parts of the book. If you're enticed into reading further, it is best to read each subsection until the discussion becomes too technical and then skip ahead until you land on a paragraph that makes sense. This approach implies accepting that I have accurately reported the science in the skipped material. I assure you that I have done so. My interpretations of the science, however, remain my own, until the concepts advanced here have been debated by the wider community of evolutionary scientists and have been either discarded or adopted. Whatever the outcome of that debate may be, I hope this book demonstrates that evolutionary science is far from a finished subject and has many more exciting discoveries in store for the rest of this new century.

Introduction: Taking a Fresh Look at the Basics of Evolution in the New Century

How does novelty arise in evolution? Innovation, not selection, is the critical issue in evolutionary change. Without variation and novelty, selection has nothing to act upon. So this book is dedicated to considering the many ways that living organisms actively change themselves. Uncovering the molecular mechanisms by which living organisms modify their genomes is a major accomplishment of late 20th Century molecular biology.

Conventional evolutionary theory made the simplifying assumption that inherited novelty was the result of chance or accident. Darwin theorized that adaptive change resulted from natural selection applied to countless random small changes over long periods of time. In Chapter 6 of *Origin of Species*, he wrote: "If it could be demonstrated that any complex organ existed, which could not possibly have been formed by numerous, successive, slight modifications, my theory would absolutely break down. But I can find out no such case" [1]. His neo-Darwinist followers took the same kind of black-box approach in the pre-DNA era by declaring all genetic change to be accidental and random with respect to biological function or need. With the discovery of DNA as a hereditary storage medium in the 1940s and early 1950s, the accidental view of change received a molecular interpretation as arising from inevitable errors in the replication process. As many professional and popular press articles attest, the accidental, stochastic nature of mutations is still the prevailing and widely accepted wisdom on the subject [2, 3].

In the context of earlier ideological debates about evolution, this insistence on randomness and accident is not surprising. It springs

1

from a determination in the 19th and 20th Centuries by biologists to reject the role of a supernatural agent in religious accounts of how diverse living organisms originated. While that determination fits with the naturalistic boundaries of science, the continued insistence on the random nature of genetic change by evolutionists should be surprising for one simple reason: empirical studies of the mutational process have inevitably discovered patterns, environmental influences, and specific biological activities at the roots of novel genetic structures and altered DNA sequences. The perceived need to reject supernatural intervention unfortunately led the pioneers of evolutionary theory to erect an *a priori* philosophical distinction between the "blind" processes of hereditary variation and all other adaptive functions. But the capacity to change is itself adaptive. Over time, conditions inevitably change, and the organisms that can best acquire novel inherited functions have the greatest potential to survive. The capacity of living organisms to alter their own heredity is undeniable. Our current ideas about evolution have to incorporate this basic fact of life.

The recognition of organically generated heritable change has its origins in classical cytogenetics, especially in the revolutionary studies of Barbara McClintock on chromosome repair and restructuring during the 1930s through the 1960s [4]. Cytogenetics is the combination of microscopic examination of chromosomes in cells with Mendelian genetic analysis. Before we knew about DNA, it was the one direct way to observe the behavior of the hereditary apparatus. The advent of molecular genetic studies, starting with bacteria in the 1950s and then expanding to all life forms with recombinant DNA technology in the 1970s, extended McClintock's insights into a universal property of microbes, plants, and animals [5] [6, 7]. Molecular analysis provided mechanistic insight into the myriad distinct ways that living cells can engineer their DNA [8]. Genome sequencing at the end of the 20th Century and the start of this one confirmed major roles played by "natural genetic engineering" in the course of evolutionary change. As we will discuss in detail in Part II, natural genetic engineering represents the ability of living cells to manipulate and restructure the DNA molecules that make up their genomes.

How Complete Is Our Understanding of Heredity?

Like the man searching for his key under the lamppost, we currently focus our thinking about heredity almost completely on DNA sequences, because our ability to read and manipulate them lies at the heart of present-day biotechnology. Nonetheless, we should never forget that not all heredity involves the transmission and interpretation of nucleotide sequences in DNA and RNA molecules. To date, all studies of genetically modified organisms have required an intact cell structure for the introduction of new genetic information by DNA or nuclear transplantation. So there is no unequivocal empirical basis for believing the frequent assertion that DNA contains all necessary hereditary information. As Rudolph Virchow articulated it in 1858, "*omnis cellula e cellula*" ("every cell comes from a cell") [10]. We still do not adequately understand the role that preexisting cell structures and organelles play in templating the formation of their descendants in progeny cells [11, 12]. In mammalian reproduction, for example, we know that both the sperm and the maternal environment contribute non-DNA factors to the fertilized egg and developing embryo [13–18]. Simple reflection makes it obvious that a properly structured egg is essential to hereditary transmission in all "higher" organisms.

Modifications of cell structure have been critical events at some of the most important stages of evolutionary innovation, and purely DNA-based explanations are insufficient to describe them in a scientifically comprehensive way. In ciliate protozoa, in the mid-20th Century, Tracy Sonneborn did pioneering work on genome-independent heredity of the cell cortex [19] [20–22]. However, his work has had few successors. Prions are now convincingly established as inherited forms of proteins encoded by the same DNA sequences as their nonprion siblings [23–29]. In addition, the importance of short- and long-term transmission of so-called "epigenetic" information contained in complexes of DNA, RNA, and protein is a burgeoning field of contemporary research with important connections to the evolutionary process [9].

Beyond these few examples, we will undoubtedly discover new aspects of cell heredity in the coming decades. It is possible that DNA-based heredity will ultimately find a more modest role in our thinking about inheritance in the course of this new century.

The contemporary concept of life forms as self-modifying beings coincides with the shift in biology from a mechanistic to informatic view of living organisms. One of the great scientific ironies of the last century is the fact that molecular biology, which its pioneers expected to provide a firm chemical and physical basis for understanding life, instead uncovered powerful sensory and communication networks essential to all vital processes, such as metabolism, growth, the cell cycle, cellular differentiation, and multicellular morphogenesis. Whenever these processes have been subjected to the most advanced types of biological analysis, the number of regulatory interactions and control molecules inevitably has grown to rival (and frequently out-number) the molecules dedicated to executing the basic biochemical and biomechanical events [30]. Paralleling the contemporaneous transformation from a largely mechanical-industrial society to a densely interconnected information-driven society, the life sciences have converged with other disciplines to focus on questions of acquiring, processing, and transmitting information to ensure the correct operation of complex vital systems.

The conceptual universe of biology inevitably underwent a radical transformation from the days of classic thinking about evolution and heredity in the 19th and 20th Centuries. That is the way of science [31]. Instead of cell and organismal properties hardwired by an all-determining genome, we now understand how cells regulate the expression, reproduction, transmission, and restructuring of their DNA molecules. The key evolutionary questions no longer center on whether we can establish relationships between different organisms. Through genome sequences, we can do that across the largest taxo-nomic distances, finding molecular features that connect the smallest microbes with the largest plants and animals. Today, instead, we endeavor to understand how complex new vital capacities arose in the course of evolution and contributed to the ability of myriad organisms to survive, proliferate, diversify, and reorganize their environment in the course of at least 3.5 billion tumultuous years of Earth history. How did evolutionary inventions help shape the biosphere and influence the nature of the organisms that inhabit it today?

We have learned enough about the diversity of existing life forms and the course of geobiological evolution to recognize that we currently see only the tip of the iceberg. At least 99% of all life forms are still without scientific description, and knowledge of the most diverse forms of life, microorganisms, is expanding daily as we discover unknown kinds in every new ecological niche we explore (including those within and upon our own bodies). But even our incomplete knowledge of the evolutionary iceberg's tip contains clues to exciting processes that were long thought (and long taught) to be impossible. The goal of this book is to acquaint you with previously "inconceivable" but currently well-documented aspects of cell biology and genomics so that you will be ready for the inevitable surprises in evolutionary science as this new century runs its course.

We will focus on how the cell rewrites its genome because that is what we know best about the sources of organic novelty. We can observe genome reorganization in real time and relate what cells do now to what the DNA record tells us has happened over the course of evolution. At both the cellular and genomic levels, the evolutionary process has clearly been one of combinatorial innovation to produce functional systems, followed by the amplification of these systems and their adaptation to novel uses.

Genomic innovations occur at many different levels of complexity. These levels cover the entire range of DNA modifications: from single nucleotide substitutions, to short strings of nucleotides comprising regulatory signals, to longer polynucleotide strings encoding functional regions ("domains") of protein molecules, through larger DNA segments encoding entire RNA or protein molecules, and finally extending to complexes of multiple coding segments and their attendant control regions. In a surprisingly large number of cases, genome analysis tells us that reorganization events have comprised whole genomes.

Because genome evolution is multilevel, amplifying, and combinatorial in nature, the end results are complex hierarchical structures with characteristic system architectures [32, 33]. Genomes are sophisticated data storage organelles integrated into the cellular and multicellular life cycles of each distinct organism. Thinking about genomes from an informatic perspective, it is apparent that systems engineering is a better metaphor for the evolutionary process than the conventional view of evolution as a selection-biased random walk through the limitless space of possible DNA configurations.

I

Sensing, Signaling, and Decision-Making in Cell Reproduction

In this part:

- How *E. coli* chooses the best sugar to eat
- Proofreading DNA replication
- DNA damage repair and mutagenesis
- Cell cycle checkpoints
- Signaling from the cell surface to the genome: pheromone response in the sexually aroused yeast cell
- The role of intercellular signals in the cell death decision
- Revisiting the central dogma of molecular biology

Living cells do not operate blindly. They continually acquire information about the external environment and monitor their internal operations. Then they use this information to guide the processes essential to survival, growth, and reproduction. Cells constantly adjust their metabolism to available nutrients, control their progress through the cell cycle to make sure that all progeny are complete at the time of division, repair damage as it occurs [34], and interact appropriately with other cells. In a multicellular context, they even undergo programmed cell death when suicide is beneficial to the entire population or to the multicellular organism as a whole [35] [36–39]. Without an elaborate sensory apparatus to pick up signals about chemicals in the environment (nutrients, poisons, signals emitted by other cells) or to keep track of intracellular events (DNA replication, organelle growth, oxidative damage), a cell's opportunity to proliferate or contribute to whole-organism development would be severely restricted. Life requires cognition at all levels [40, 41].

Molecular biology has successfully documented many examples of cellular information acquisition, transmission, and processing, some of which will be described here in Part I. But we should remember that there is not yet any deep theoretical understanding of the basic principles of cell informatics. Developing that understanding is a major biological research goal of the present century. The best we can do right now is to recognize that cells utilize many kinds of molecular interactions to process information and execute appropriate decisions [42, 43].

Because the interactions in any cell process invariably grow more complex and involve more molecules as we investigate them in greater detail, most biologists agree that we are now in the *systems biology* era of research [44–46]. Although this term is subject to various interpretations, a widespread view is that systems biology implies understanding how groups of molecules work coordinately (as a system) to achieve some useful function dependent upon conditions. Gone is the atomistic view that molecules act independently and automatically.

How *E. Coli* Chooses the Best Sugar to Eat

One of the most important chapters in the history of molecular genetics began in Nazi-occupied Paris in 1942. That was when Jacques Monod (who was also a leader of the Paris Resistance) published his doctoral thesis on measuring the growth of bacterial cultures [47]. Monod determined quantitatively the effects of providing different nutrients in various amounts on the speed and extent of bacterial growth. He established that some sugars (such as glucose) are better than others (such as lactose) at powering the *rate* of bacterial growth (the increase in both total mass and cell number per hour). But he did not find a significant difference in final growth *yield* per unit of sugar provided.

As Monod explored these quantitative relationships, he happened to grow the bacteria on mixtures of more than one sugar and found a surprising result. When bacteria grow on a mixture of high- and low-growth-rate sugars, the growth process occurs in two distinct phases: a more rapid one followed by a pause before the bacteria begin a slower growth phase. By the simple but elegant procedure of mixing the sugars in various proportions, he proved that the bacteria completely consumed the preferred sugar before starting to consume the less rapidly utilized sugar. This was a striking discovery. Monod called the two-stage growth process *diauxy*, which means "double growth." It implied that the bacteria could distinguish between the two sugars and adjust their metabolism to utilize the preferred sugar first until it was completely consumed. Then, when the preferred sugar was no longer available, the bacteria took a certain time to readjust their metabolism and subsequently began to utilize the less preferred sugar. How did all of this occur?

Monod, his colleagues, and their followers spent the next three decades investigating the consequences of this biphasic growth phenomenon in the bacterium *E. coli*, which had become the chief model organism in the new science of molecular genetics [48, 49]. These investigations led to a series of fundamental discoveries and realizations that transformed our understanding of how cells organize information in their DNA and access that information in response to a changing environment. Accordingly, let us examine some of the details behind this molecular recognition process and then discuss their more general implications.

Control of bacterial sugar metabolism, especially of the choice by *E. coli* between glucose and lactose, is one of the most basic and fully described cell regulatory systems known **[50, 51]** [52]. Even so, the number of molecular components involved is sufficient to require some concentration to master. The details are summarized in Appendix I.1 online. The non-technical reader might be tempted to skip the details, but the only way to appreciate biological information processing fully is to dig in and see how the circuits operate in a logical, Boolean fashion. These logical circuits are the products of evolution. They also illustrate principles that will prove critical to formulating a 21st Century view of the evolutionary process.

From examining the *relatively* simple regulatory circuit controlling expression of the *E. coli lac* operon,[1] molecular biologists have uncovered at least five general principles of cellular information processing and communication with the genome:

---

1 An *operon* is a coordinately expressed group of adjacent coding sequences under the control of a common regulatory site [53].

- There is no Cartesian dualism in the *E. coli* (or any other) cell. In other words, no dedicated information molecules exist separately from operation molecules. All classes of molecule (proteins, nucleic acids, small molecules) participate in sensing, information transfer, and information processing, and many of them perform other functions as well (such as transport and catalysis).

- Information is transferred from cell surface or intracellular sensors to the genome using relays of proteins, *second messengers*, and DNA-binding proteins [54–56]. The term *second messenger* is applied to small molecules, such as the ATP derivative cyclic adenosine monophosphate (cAMP) and the inducing lactose metabolite allolactose, that serve as diffusible chemical signals and carry information to the macromolecules that bind them [57]. Note that we can view molecules such as cAMP as part of the cell's symbolic chemical lexicon because there is no direct structural relationship between it and the metabolic information it represents. There are a number of attempts to describe cellular information processing from a semiotic or linguistic perspective [58, 59].

- Protein-DNA recognition often occurs at special recognition sites (such as *lacO*, *lacP*, and *crp*, as defined in Appendix I.1 online). The description of the first of these recognition sites, *lacO*, in 1961 was revolutionary in our understanding of the genome because such sites are fundamentally different in nature and function from the protein coding sequences conventionally believed to constitute "genes" [60]. Recognition sites format the DNA for interactions in many processes besides transcription, such as DNA compaction, DNA replication, DNA transmission to daughter cells, and DNA restructuring [30, 61, 62]. In some aspects, genome formatting can usefully be compared to data file formatting in computer systems because they are essential in both cases to accurate utilization of stored information [63]. The realization that all aspects of genome function involve formatting elements is one of many reasons that the term "gene" has become impossible to define rigorously. Changes in formatting are as important as changes in coding capacity in altering genome behavior and expression and thus organic phenotype.

- DNA binding proteins and their cognate formatting signals operate in a combinatorial and cooperative manner **[64]**. This is exemplified in the *lac* operon by the interaction between RNA polymerase and CRP at the *lacP* and *crp* sites to stimulate transcription [65]. In addition, the *lac* operon contains three different *lacO* sequences, and proper repression by LacI requires cooperative binding at two of the three sites [66]. Changing the spacings between these formatting signals significantly alters the regulatory interactions [67]. In other systems, particularly in plants and animals, the combinatorial and cooperative nature of protein-DNA interactions allows complex formatting regions (often called *cis-regulatory modules* or CRMs) to be constructed from a finite number of different recognition sequences [68].

- Proteins operate as conditional microprocessors in regulatory circuits. They behave differently depending on their interactions with other proteins or molecules. The effect of the LacI repressor binding the allolactose inducer and consequently losing its ability to interact repressively with the *lacO* DNA sequence is a classic example of what are called *allosteric* transitions in proteins (changes in shape due to ligand binding that alter functional properties) **[69, 70]** [48, 71, 72]. In essence, the LacI repressor is a microprocessor modulating genome expression in response to the inducer (a second messenger) because inducer binding appears to exclude operator binding. The cAMP-CRP interaction illustrates the same principle but in the opposite direction (second messenger binding stimulates DNA attachment) [65, 73]. Another variant of the same principle is the effect of phosphorylation on the glucose transport protein that enables it to stimulate cAMP synthesis [74]. There are virtually limitless ways of reversibly altering protein structure and interactions with cellular molecules, including DNA and other proteins. Thus, it should not be difficult to envisage the construction of enormously complex computational circuits in living cells that use allosteric protein microprocessors to regulate genome expression [43, 75, 76].

Proofreading DNA Replication

One of the central ideas in the conventional approach to evolutionary change is that DNA alterations are accidental—they arise from unavoidable errors in the replication process or from physicochemical damage to DNA molecules. Active cell processes that ensure the accuracy of replication or repair DNA damage are, in the conventional view, taken for granted in genome maintenance. However, the fact that cells actively maintain genome integrity during normal growth is a key aspect of cell control over its major information storage organelle. As more than five decades of research have established, DNA proofreading and repair systems are central to the yin and yang of cell management of genome structure: conservation in times of successful growth as compared to active restructuring in times of stress. Intracellular sensory processes are key to both complementary aspects of genome maintenance.

The *E. coli* cell reproduces its DNA with remarkable precision (less than one mistake for every billion (10^9) new nucleotides incorporated) and at surprisingly high speed. The *E. coli* cell duplicates its 4.6 MB genome in 40 minutes (about 2,000 nucleotides per second), independently of the cell division time [77]. The extraordinarily low error frequency results from monitoring the results of the polymerization process and correcting incorporation mistakes after the fact, not from the inherent precision of the replication apparatus. The DNA polymerase that incorporates nucleotides itself has an intrinsic precision of about one mistake for every 100,000 (10^5) nucleotides [78]. Although this is impressive when compared to any man-made manufacturing process, the polymerase alone is at least four orders of magnitude less accurate than the final replication result. Ultimate precision is achieved by two separate stages of sensory-based proofreading:

- The first stage occurs during polymerization. When an incorrect nucleotide has been incorporated into the new growing DNA strand, mispairing between the new and old strands distorts the structure of the growing double helix. The polymerase senses the distortion and interrupts polymerization. While polymerization is halted, another activity of the replication apparatus removes the incorrect base from the end of the new strand, relieves the distortion to the double helix, and allows polymerization to resume, replacing the incorrectly inserted nucleotide. In this process, known as *exonuclease proofreading*, the polymerase itself serves as the sensor that detects mistakes and activates the correction functions **[79, 80]**. Exonuclease proofreading increases replication accuracy 100- to 1,000-fold (two to three orders of magnitude) [78, 81, 82].

- The second stage of replication proofreading operates to detect and remove incorporation errors that escape exonuclease proofreading. This process is called *mismatch repair* and utilizes three different proteins dedicated to postreplication proofreading. The Mut acronym in their names indicates that loss of each protein leads to a *mutator* phenotype. MutS scans the newly replicated DNA and binds to regions where the double helix has been distorted by a mismatch. MutL recognizes the mismatch-bound MutS and connects it with the third protein, called MutH, which cleaves the newly synthesized strand on both sides of the mismatch. These strand cleavages allow other cell proteins to remove (*excise*) a length of newly synthesized DNA carrying the misincorporated nucleotide and replace it with a newly synthesized, error-free DNA strand. Special features of the newly synthesized DNA allow the MutH endonuclease to discriminate it from the old strand, thus ensuring that misincorporations are corrected rather than fixed into both strands of the genome **[79]**. The efficiency of the mismatch repair system is about 99%, increasing replication accuracy by a further 100-fold to its amazing final precision of less than one mistake per billion incorporations [83–86].

In mismatch repair, we observe several examples of molecular sensing:

- MutS senses DNA helix distortions due to mismatch.
- MutL recognizes MutS once it is bound to mismatched DNA and serves as a molecular "matchmaker" to bring MutS into contact with the MutH endonuclease [87].
- MutH discriminates between the new and old strands of recently replicated DNA. In *E. coli*, this strand discrimination depends on the fact that the older parental strain is methylated while the new strand is unmethylated **[79]**.

Note that all three mismatch repair proteins display allosteric micro-processor-like behavior; they change the nature of their interaction with one particular molecular partner based on whether a prior inter-action with another partner has already occurred.

The two-stage replication proofreading system in *E. coli* and other bacteria has more complex analogues in nucleated eukaryotic organisms, from yeast to plants and animals.[2] In these organisms, the molecules of the damage sensory apparatus have been amplified and refined. Eukaryotes contain a plethora of different DNA polymerases and proofreading exonucleases [88–91]. Eukaryotes also possess multiple proteins resembling MutS and MutH that operate in various combinations to recognize specific types of mismatches and direct the error excision and replacement process. We know that these systems play a critical role in our own cells because defects in mismatch repair result in an inherited tendency to develop colon cancer [83, 86, 92].

We can think of this two-level proofreading process as equivalent to a quality-control system in human manufacturing. Like human quality-control systems, it is based on surveillance and correction (cognitive processes) rather than mechanical precision. The multistep nature of proofreading is typical of many control processes in cells, where final precision is achieved by a sequence of two or more inter-actions that are each themselves inherently less precise. In this regard, the most applicable cybernetic models are *fuzzy logic* control systems. In such systems, accurate regulation occurs by overlaying multiple imprecise ("fuzzy") feedback controls arranged so that each successive event results in greater precision **[93]** [94, 95].

DNA Damage Repair and Mutagenesis

Another common misperception in many conventional discussions of genomic change is that cells cannot avoid the automatic production of mutations in response to DNA-damaging agents such as UV radiation or mutagenic chemicals. This misperception results from ignorance about the sophisticated apparatus that even the smallest cells possess

2 *Eukaryotic*—literally "true kernel"—organisms have a defined nucleus containing chromosomes surrounded by a membrane. Bacteria, such as *E. coli*, are *prokaryotic* organisms that do not have a defined nucleus.

to repair genome damage and a failure to appreciate the power of cellular genome surveillance and response regimes. The subject of DNA repair has become a veritable research industry because of its relationship to carcinogenesis [83]. To keep our discussion to a manageable length, we will restrict it to the effects of UV radiation and related chemical agents in bacteria.

The first clear indications that cells play an active role in repairing DNA damage and creating mutations in response to UV and other mutagenic agents came in the early years of molecular genetics. Many of the experiments were done with *E. coli* and other bacteria, where it was found that the lethal and mutagenic effects of UV irradiation were reduced by keeping the bacteria for various periods of time in nongrowing conditions [96, 97]. This observation indicated that the bacteria could remove damage from their genomes and not suffer either death or mutation if they were not actively replicating their DNA. A further experiment indicated that this repair capacity was actually the result of a cell response induced by the damage itself. If the bacteria were irradiated and then treated with a nonlethal antibiotic that prevented new proteins from being synthesized, most of the repair capacity was eliminated. This result meant that the cells expressed newly induced genomic DNA repair information after irradiation **[98]**.

The most direct indications of an inducible cellular role in DNA repair and mutagenesis came from ingenious experiments in the 1950s by Jean Weigle, a Swiss physicist turned molecular geneticist. Weigle utilized an *E. coli* virus (or *bacteriophage*) named λ. He took advantage of the fact that isolated virus particles provided a source of target DNA independent of the cells in which the virus had to reproduce [99]. By irradiating either λ and/or the cells to be infected, he could distinguish between the effects of UV radiation on the DNA (in the virus particles) and on the cell in which the DNA had to replicate. Weigle found that irradiated cells were much better at repairing lethal UV damage than unirradiated cells, thereby confirming the inducible nature of the cellular repair process. Surprisingly, he also found that irradiated cells produced more λ mutations than unirradiated cells, even when infected with *unirradiated* virus. This result, which came to be known as *Weigle mutagenesis*, demonstrated that UV radiation induced a mutagenic capacity in the bacteria, which was active even on "untargeted" DNA that was not subjected to UV damage [100–104].

Molecular genetic analysis ultimately revealed that the inducible UV repair and mutagenesis capabilities of *E. coli* and other bacteria form part of a complex, highly orchestrated whole-cell response to DNA damage labeled the *SOS response* by Evelyn Witkin, a pioneer of repair and mutagenesis studies **[98]** [105–107]. The SOS response involves two kinds of repair systems:

- A precise repair process that removes the UV-damaged DNA and does not introduce mutations. This "error-free" process operates very much like the mismatch repair proofreading system, except that the sensor protein recognizes the characteristic chemistry of UV damage in DNA rather than helix distortions. It is called *excision repair* because the result of damage sensing, as in mismatch correction, leads to excision of a section of the damaged DNA strand [108–110].

- A mutagenic repair process that involves the synthesis of specialized "error-prone" DNA polymerases, which can replicate DNA that carries unrepaired damage. Without these specialized polymerases, mutations do not occur in response to DNA damage. Instead, the cells or molecules that cannot remove or replicate past the damage are simply doomed and produce no mutant progeny [111–114].

The fact that cells lacking particular biochemical functions are not subject to induced mutagenesis is a basic piece of evidence for the inherently biological nature of genetic change. In fact, biochemical activities determine the kinds of genetic changes that can occur in response to DNA damage. A rather poorly known but elegant and ingenious set of experiments has demonstrated that different types of localized "point" mutations can each be attributed to the action of particular mutagenic polymerases [115].

In addition to overlapping error-free and mutagenic repair processes, the SOS response involves the synthesis of proteins that promote homologous recombination, halt cell division, alter cell metabolism, inhibit normal DNA replication, and stimulate exit from the SOS state after repair is complete [106, 116, 117]. Homologous recombination (HR) is the process in which two DNA molecules that share the same sequences (that have homology) undergo physical and genetic exchange of material [118]. The SOS system depends on a multifunctional sensor protein called RecA [119]. RecA received its name from A. John Clark, who discovered the central role it plays in the process of homologous recombination, which is a key mechanism for repairing double-stranded breaks (DSBs) in DNA molecules [120–123].

RecA protein forms multicopy filaments on exposed regions of single-stranded (SS) DNA, which accumulate following certain kinds of DNA damage, such as a collapse or blockage of the replication machinery [124]. The RecA-SS DNA filament remarkably stimulates two quite different biochemical events:

- The base pairing of complementary DNA strands to form a duplex segment, a key step in homologous recombination [118] [125–127]

- The cleavage of proteins, including a repressor protein that regulates expression of other SOS functions [128–132]

It is the protein-cleavage activity that provides the sensory connection between DNA damage, accumulation of SS DNA in the cell, and consequent activation of the complex SOS response. The RecA-SS DNA filament induces synthesis of all the diverse repair and additional proteins that execute the full range of SOS functions.

While this discussion of inducible repair and mutagenesis in response to DNA damage has focused on bacteria for the sake of relative simplicity, far more elaborate DNA damage response systems are well studied in eukaryotic cells. Because of the clear connection to cancer, extraordinary attention has been paid to these systems and how they affect genome change in response to chemical damage. Rather than belabor the points already illustrated with bacteria, we will leave discussion of the eukaryotic DNA damage response to the next section on cell cycle regulation.

Testing for Mutagens and Carcinogens Turned Out to Be a Test for Mutator Protein Synthesis Rather Than a Direct Test for DNA Damage

One way to emphasize that mutagens affect the genome through a cellular biochemical response rather than simply by direct interference with replication is to look at the well-known bacteria-based *Ames test* for mutagens and carcinogens [133]. In the early 1970s, the bacterial geneticist Bruce Ames had the clever idea that it would be far more rapid and economical to check chemicals for carcinogenicity using bacteria as test organisms rather than mice. Ames had an extensive collection of mutations affecting synthesis of the amino acid histidine in *Salmonella typhimurium*, an intestinal pathogen closely related to *E. coli*. The test consisted of placing the compound in question on a Petri dish in contact with a layer of *Salmonella* mutants unable to grow for lack of histidine and see if the chemical induced reverse mutations that restored the ability to synthesize histidine (and thus form visible colonies). Initial versions of the test gave disappointing results because the *Salmonella* SOS response lacked significant levels of mutagenic polymerase activities. When these SOS-inducible polymerase activities were introduced to the test strains, the sensitivity of the screening process increased greatly and provided a useful method for determining the *genotoxicity* of chemical compounds [134, 135]. What the requirement for inducible mutagenic polymerases demonstrated was that the Ames test actually determined SOS activation rather than DNA damage. This point was confirmed when an even more rapid and sensitive mutagen/carcinogen test was developed that directly measured induction of SOS expression without waiting for mutagenesis to occur [136].

Cell Cycle Checkpoints

One of the most important components of the SOS response is the SulA protein, which blocks the cell division process [137, 138]. By doing so, SulA prevents the formation of progeny cells receiving a damaged or incomplete genome. Because SulA is unstable and has a very short half-life, cell division resumes once the damage has been repaired and the SOS repressor recovers to block further SulA synthesis. Through its action to halt the bacterial cell cycle until DNA repair is complete, SulA serves what has come to be known as a *checkpoint* function. Checkpoints are surveillance-dependent controls on the various steps in the indescribably complex business of cell reproduction; they guarantee that the entire process does not move forward until all the preliminary steps, such as ensuring genome integrity, have been completed [139].

The first explicit articulation of the checkpoint concept by Weinert and Hartwell in 1988 involved a yeast function, RAD9, analogous in some ways to SulA [140]. RAD9 is also involved in blocking cell division until radiation-induced damage had been repaired. The name indicates that loss of the RAD9 protein makes budding yeast cells hypersensitive to radiation damage.[3] Unlike the direct action of SulA on cell division, RAD9 works as part of an intricate network of proteins that constantly monitor the status of the yeast genome and connect the information they acquire to another network that regulates progression through the tightly controlled eukaryotic cell cycle [141].

In their normal vegetative reproduction, eukaryotic cells pass through several stages in what is called the *mitotic* cell cycle **[142]**:

- G1, first growth phase.
- S, DNA replication (synthesis) phase.
- G2, second growth phase.
- M, mitosis (division) phase.

Like all biological processes, cell cycle progression is closely monitored and regulated [143, 144]. The basic principle of checkpoint control is that information about delays, errors, or damage in genome replication and other aspects of cell development (such as daughter bud formation) can be transmitted to the molecular reactions that control transitions from one cell cycle stage to another [145–147]. The checkpoint signals bring the cell cycle to a halt at any point prior to cell separation and prevent the formation of inviable daughter cells.

---

3 Budding yeast is called *Saccharomyces cerevisiae*, or brewer's yeast, because it is the organism that ferments most beers.

The published diagrams illustrating the biochemical control of cell cycle progression grow more complex every year—and more interconnected with all aspects of cell proliferation (http://www.sabiosciences.com/pathwaycentral.php?application=CELLCY). In the case of genome surveillance, there are separate but partially overlapping checkpoint systems: one for damage in G1 before genome doubling, one for replication problems in S, one for damage in G2 after genome doubling, and one for abnormalities in behavior and alignment of the duplicated chromosomes on the mitotic spindle apparatus in M [148].

As previously exemplified by the fidelity of DNA replication, cognitive checkpoint control rather than mechanical precision ensures the reliability of eukaryotic cell division. Consideration of the *spindle checkpoint* in the M phase of the cell cycle illustrates this argument [149–151]. It is sometimes claimed that chromosome distribution at mitosis is random because there is no way to predict which of the two chromosome copies will end up in a particular daughter cell. But the fact that one and only one chromosome copy goes to each daughter actually makes the process highly nonrandom. If chromosome distribution were truly random, only 50% of cell divisions would produce progeny with a single copy of each duplicated chromosome in each daughter cell. For an organism such as S. *cerevisiae*, with 16 chromosomes, random chromosome distribution to daughter cells would result in only $1/2^{16} < 1/32,000$ divisions producing two progeny cells with equal complete genomes.

The job of the spindle apparatus is to make sure that each daughter cell receives a single copy of each duplicated chromosome. This guarantees that each daughter cell has a complete genome. We now have a fairly good idea of how this checkpoint operates. The mitotic chromosome separation apparatus involves attaching microtubule filaments that radiate from the spindle poles to *kinetochore* structures built on the centromere of each duplicated chromosome, which remains attached to its sibling copy [152] [153]. If each sibling kinetochore is attached to microtubules from the opposite pole, the duplicate chromosomes move appropriately to opposite spindle poles and thus end up in different daughter cells. Correct microtubule association generates an equal tension on the sibling chromosomes, and no checkpoint signaling occurs. However, if a chromosome is improperly attached, or if both chromosomes are attached to the same spindle pole, equal tension is not achieved, and a checkpoint signal prevents chromosome movement toward the spindle poles [154]. The end result of this monitoring-communication system is that chromosome separation, telophase, and cytokinesis (the completion of eukaryotic cell division) do not occur until all chromosome pairs are properly aligned on the spindle. This delay process ensures that proper genome transmission to each daughter cell occurs with a high degree of certainty (typically > 99.99%).

Signaling from the Cell Surface to the Genome: Pheromone Response in the Sexually Aroused Yeast Cell

If we stay with budding yeast, we can discuss a classic and relatively simple example of intercellular signaling that illustrates the ability of one cell to communicate with the genome and cell cycle control apparatus of another cell. S. cerevisiae cells have a sexual cycle that involves a fascinating courtship process between cells of opposite sexes, or mating types: **a** and α [155, 156]. Courtship begins as cells of opposite types approach and emit **a** and α mating pheromones (in yeast, these are short protein molecules decorated with lipids) [157]. Haploid yeast cells contain receptors for pheromones emitted by cells of the opposite mating type so that they can detect them and respond appropriately. The response includes stopping normal growth in the G1 phase (with an unduplicated haploid genome) and developing "shmoo-like" cell outgrowths that extend in the direction of the opposite cell type [158]. These outgrowths eventually merge to form a joint cell with two haploid nuclei that subsequently fuse to produce the nucleus of an **a**/α diploid cell.

This elaborate courtship and mating process involves each cell's inducing three major changes in its partner's internal functions:

- G1 arrest of the cell cycle (a halt to DNA replication).
- Expression of the mating-specific functions needed for shmoo-like growth followed by cell and nuclear fusion.
- Oriented cell morphogenesis toward the pheromone-emitting partner.

Watching the courtship and mating behavior of these eukaryotic microbes through a microscope generates admiration for their capacities to signal, sense, and exert control over the genome and morphogenesis of another yeast. A cell of one mating type can discriminate between two cells of the opposite mating type [159], so it is more than a completely automatic process. Because the α-pheromone produced by α cells has proved easier to isolate and use for laboratory experiments, the molecular biology of pheromone response has been examined in the **a** cells that respond to it. Further details of the signaling process that connects the surface receptor to the genome and cell-cycle control apparatus and a series of micrographs illustrating oriented growth and fusion are presented online in Appendix I.2.

There are two particularly significant facts about the *S. cerevisiae* mating pheromone response:

- It provides a concrete example of how one cell can communicate through well-defined molecular events with the genome of another cell. This example shows that we cannot consider the genome in any way isolated from the outside world; it is a fully informed cell organelle that works dynamically in response to a wide range of organic and inorganic inputs.

- The yeast pheromone response system utilizes sophisticated and complex sensory and signaling components (such as a G protein-coupled receptor) that researchers have encountered repeatedly in many quite distant organisms **[160]**.

G protein-coupled receptors linked to MAPK kinase signaling cascades are used throughout eukaryotic cell biology. The PubMed database of biological articles contains more than 231,250 references for "G protein coupled receptor" and more than 42,373 for "MAPK kinases," many of which deal with human biology. The repeated use of these molecular sensing and signaling systems teaches us that important evolutionary inventions are maintained and reused over very long periods in evolutionary history. We will examine further evidence for this kind of evolutionary conservation in Part III. Why we should expect evolutionary inventions to be reused and adapted to new functions will be considered from a more theoretical perspective in Part IV.

The Role of Intercellular Signals in the Cell Death Decision

One of the strongest indications that cells are actively in control of their destinies has been the discovery of the processes that determine when a cell dies. Although our initial assumption is generally that cells die when they receive an irreparable trauma or accumulate an overwhelming burden of defects with age (a process called *necrosis*), it turns out that a significant (perhaps overwhelming) proportion of cell deaths result from the activation of biochemical routines that bring about an orderly process of cellular disassembly known by the terms *programmed cell death* and *apoptosis* (from the Greek, meaning "falling away") **[35]** [39, 161, 162].

Apoptosis was first recognized as distinct from trauma-induced necrosis in the 19th Century, but its wider significance was only recognized in the 1960s and 1970s, when studies of development in nematode worms revealed regular patterns of cell death in the course of embryonic morphogenesis. In addition, the worm studies identified some of the inherited factors responsible for this reproducible cell death pattern, demonstrating that it was an active, regulated function of the organism [36]. Subsequent research has expanded our knowledge of the many cascading components that govern animal cell apoptosis [163].

Today, we recognize that nematode-like apoptosis and other forms of programmed cell death occur in a wide range of organisms for various purposes. In our own embryonic development, for example, apoptosis removes cells connecting emerging fingers and toes so that they can move independently [164]. Apoptosis is also a possible outcome of the human DNA damage response [165, 166], where it protects the whole organism from the formation of aberrant precancer cells. In plants, cell death is induced as a defense against infection, where it is called the *hypersensitive response* [37, 167]. After bacterial or viral infection, the infected plant cell emits signals so that it is quickly surrounded by a zone of dying cells, which provides a barrier to prevent the reproduction and spread of the infecting pathogen. There are even programmed cell death functions in bacteria, where they maintain genetic stability and ensure survival of a proportion of the cells in multicellular populations [38]. We know that bacterial apoptosis is a multicellular activity in bacteria because it involves a special intercellular signaling molecule [168, 169].

In all cases of programmed cell death, intercellular signaling plays a key role [170]. Mammalian cells have *death receptors* for physiological signaling proteins, such as tumor necrosis factor (TNF), which can activate the apoptosis cascade [171, 172]. But the apoptosis response is not hardwired. Mammalian cells also have receptors for growth- and survival-promoting factors, such as insulin-like growth factor (IGF) and epidermal growth factor (EGF). When these receptors are bound, the intracellular signaling networks activate molecules that can block transmission of the apoptosis response [173, 174]. Thus, each cell has the potential to make a signal-dependent life-or-death decision [175–178]. Experiments have confirmed that cells respond to lethal events such as DNA damage differently according to the intercellular signaling factors and extracellular matrix present in their environment [179]. For example, cells deprived of molecules such as IGF and EGF display a far higher apoptotic response to radiation than do cells supplied with the growth factors [180, 181].

Revisiting the Central Dogma of Molecular Biology

The selected cases just described are examples where molecular biology has identified specific components of cell sensing, information transfer, and decision-making processes. In other words, we have numerous precise molecular descriptions of cell cognition, which range all the way from bacterial nutrition to mammalian cell biology and development. The cognitive, informatic view of how living cells operate and utilize their genomes is radically different from the genetic determinism perspective articulated most succinctly, in the last century, by Francis Crick's famous "Central Dogma of Molecular Biology." So it is appropriate to direct our attention to evaluating the validity of Crick's formulation in light of 21st Century knowledge.

Crick first published the Central Dogma idea in 1958 to articulate the growing knowledge about the molecular basis of protein synthesis. The prevailing idea then was that DNA determined heredity by encoding protein structure, and it was assumed that proteins acted to determine the phenotype of the cell and organism [182]. Crick postulated two linear information flows based on nucleotide sequence coding: DNA —> DNA during replication and DNA —> RNA —> protein during protein synthesis. In 1970, Crick revised his unidirectional formulation in light of Mizutani and Temin's then-recent discovery of *reverse transcriptase* activity that could copy RNA back into DNA [183] [184, 185]. Crick allowed an extra arrow from RNA to DNA in his scheme, but he wrote that transfers of information from protein to nucleic acid or from protein to protein were unacceptable: "...sequence information cannot be transferred from protein to either protein or nucleic acid" and "the discovery of just one type of present day cell which could carry out any of the three unknown transfers (protein —> DNA, protein —> RNA, protein —> protein) would shake the whole intellectual basis of molecular biology..." [185].

Clearly, in 1970 Crick held a Cartesian dualist's view of molecular information transfer, in which nucleic acids contained the coded information and proteins executed the encoded instructions. A contemporary version of this DNA-centric view can be seen in the article "Deciphering the Code of Life" [186]. Today, we know about many examples where proteins modify sequence information in DNA (such as SOS mutagenesis), in RNA (splicing and other types of posttranscriptional processing), and in other proteins (proteolytic cleavages, peptide excisions, and peptide attachments). We also have far deeper insight into the many ways that proteins and other cell molecules left out of Crick's scheme (second messengers, membranes, and noncoding RNAs) influence the structure, expression, and modification of both DNA in the genome and RNA transcripts (see Table I.1).

From the kind of information transactions listed in Table I.1, it seems that "the intellectual foundations of molecular biology" have indeed been shaken—and shaken hard. The purpose of Part I of this book is to introduce you to a small number of the many cases where molecular biology has taken us into new conceptual territory. In particular, the first proposition of the contemporary view of cell information processing, DNA + 0 —> 0, makes the point that DNA cannot do anything or direct anything by itself; it must interact with other cell molecules [187]. So all genome action is subject to the inputs and information-processing networks we know to operate in living cells.

Part II describes how the new conceptual landscape leads us to think about a read-write (RW) genome, replacing the traditional evolutionists' read-only memory (ROM) device subject to change by accidents and errors.

Table I.1 Changing Views of Intracellular Molecular Information Transfer

1970 [185]

(DNA —> 2X DNA) —> RNA —> protein —> phenotype

2009 [188]

DNA + 0 —> 0

DNA + protein + ncRNA —> chromatin/epigenetic markings (epigenotype)

Chromatin + protein + ncRNA —> DNA replication, chromatin maintenance/reconstitution

Protein + RNA + lipids + small molecules —> signal transduction

Signals + chromatin + protein —> RNA (primary transcript)

RNA + protein + ncRNA —> RNA (processed transcript)

RNA + protein + ncRNA —> protein (primary translation product)

Signals + chromatin + proteins + ncRNA + lipids —> nuclear/nucleoid localization

Protein + nucleotides + Ac-CoA + SAM + sugars + lipids —> processed and decorated protein

DNA + protein —> new DNA sequence (mutator polymerases, terminal transferases)

Chromatin + protein —> new DNA structure (DNA-based rearrangements)

RNA + protein + chromatin —> new DNA structure and sequence (retrotransposition, retroduction, retrohoming, diversity-generating retroelements)

Protein + ncRNA + chromatin + signals + other molecules + structures <—> phenotype & genotype & epigenotype

The Genome as a Read-Write (RW) Storage System

In this part:

- Genome formatting for proper access to stored information
- Genome compaction, chromatin formatting, and epigenetic regulation
- Genome formatting for replication, localization, and transmission to daughter cells
- Distinct classes of DNA in the genome
- The molecular mechanisms of natural genetic engineering
- Natural genetic engineering as part of the normal life cycle
- Cellular regulation of natural genetic engineering
- Targeting of natural genetic engineering within the genome
- Reviewing what cells can do to rewrite their genomes over time

Today's molecular biologists have gained detailed knowledge about regulation of the cell cycle, genome transmission at cell division, transcriptional and epigenetic regulation of genome expression, and natural genetic engineering. In acquiring this knowledge, they have entered an intellectual realm far more compatible with the informatic, cognitive perspective outlined in Part I of this book than with the conventional mechanistic viewpoint held since the early 20th Century by classical geneticists, by formulators of the Modern Evolutionary Synthesis of Darwinism and Mendelism, and by the early pioneers of molecular biology [189–194].

Stated in terms of an electronic metaphor, the view of traditional genetics and conventional evolutionary theory is that the genome is a read-only memory (ROM) system subject to change by stochastic damage and copying errors. For over six decades, however, an increasingly prevalent alternative view has gained prominence. The alternative view has its basis in cytogenetic and molecular evidence. This distinct perspective treats the genome as a read-write (RW) memory system subject to nonrandom change by dedicated cell functions [4, 40, 62, 187, 188, 195–226]. The radical difference between the ROM and RW views of genomic information storage is basic to a 21st Century understanding of all aspects of genome action in living cells. Cell-mediated inscriptions on the genome occur at all time scales, ranging from the single cell cycle to evolutionary epochs [62, 187].

To appreciate this paradigm shift in genomic concepts, we will examine several modern lessons about how a data storage/memory organelle operates in living cells. Although it's useful in many ways, the computer metaphor can be misleading if we do not put genome function within a proper biological context. In particular, it is often asserted that the basic function of a genome is simply to hold data files, or "genes," which determine the sequence structures of RNA and protein molecules, which in turn determine the organism's properties. This view was often summarized in the "beads on a string" metaphor for the linking of genes on chromosomes. The 1948 article in the online suggested readings by George Beadle, author of the "one gene, one enzyme" hypothesis, explicitly articulates this point of view [227]. But data file maintenance is an inadequate and incomplete description of genome function. Within the living, dividing cell, there are other requirements that the genome must meet and integrate into its functional organization. We can distinguish at least seven distinct but interrelated genomic functions essential for survival, reproduction, and evolution:

1. DNA condensation and packaging in chromatin
2. Correctly positioning DNA-chromatin complexes through the cell cycle
3. DNA replication once per cell cycle
4. Proofreading and repair

5. Ensuring accurate transmission of replicated genomes at cell division

6. Making stored data accessible to the transcription apparatus at the right time and place

7. Genome restructuring when appropriate

In all organisms, functions 1 through 6 are critical for normal reproduction, and (as you will see later) quite a few organisms also require function 7 during their normal life cycles. We humans, for instance, could not survive if our lymphocytes (immune system cells) were incapable of restructuring certain regions of their genomes to generate the essential diversity of antibodies needed for adaptive immunity. In addition, function 7 is essential for evolutionary change. This part of the book devotes considerable attention to discussing the numerous molecular modalities of genome restructuring discovered since the double helix structure of DNA was described in 1953.

Why and How to Avoid Using the Term "Gene"

Throughout the book, the term "gene" appears in quotation marks to indicate its hypothetical nature. This term has no rigorous and consistent definition. It has been used to designate countless different features of genome organization. In other words, the use of "gene" gives the false impression of specifying a definite entity when, in fact, it can mean any number of different genomic components.[1]

In some cases, such as when "gene" is used to indicate a continuous human DNA sequence encoding a specific protein, it actually means something that generally has no real existence in nature. This is because the genomic DNA coding regions usually comprise several separated exons and are only joined at the level of the RNA transcript. In place of "gene," therefore, the term *coding sequence* indicates DNA regions that determine protein primary structure,

---

1 See Natalie Angier's 2008 *New York Times* article "Scientists and Philosophers Find That 'Gene' Has a Multitude of Meanings" at http://www.nytimes.com/2008/11/11/science/11angi.html?_r=3&sq=RNA%20genes&st=cse&scp=2&pagewanted=print.

in keeping with the contemporary use of *CDS* to indicate protein-encoding regions in genome sequence annotations. Moreover, the term *genetic locus* is preferable, because it indicates a discrete identifiable region of the genome, including signals formatting transcription and post-transcriptional processing, that encodes one or more messenger RNA molecules, which in turn encode one or more specific protein products [228]. The reason for this preference is that a genetic locus can be defined operationally by either genetic or molecular analysis, and the two-word phrase does not carry the same theoretical and confusing implications as the multifarious term "gene" [229].

Genome Formatting for Proper Access to Stored Information

In Part I, we discussed the conceptual revolution in genetics due to the discovery of protein recognition signals used to control transcription of the *lac* operon and other coding sequences in both bacterial and eukaryotic genomes. The term *coding sequence* indicates the DNA sequence that determines the primary structure of an RNA or protein product. Over the last three decades of the 20th Century, our understanding of genome formatting for transcriptional regulation grew tremendously, and it continues to expand vigorously [64] [68, 230, 231]. We are still unraveling the sophisticated potentials of transcriptional formatting for executing complex regulatory tasks. References to some of the major principles to emerge from this ongoing exercise are summarized in Appendix II.1 online.

The take-home lesson from studies of transcriptional regulation is that genomes contain the basic formatting elements necessary to build elaborate circuits that control expression of coding sequences in complex ways. These circuits involve sequence motifs (relatively short DNA sequences that serve as recognition signals for interaction with proteins or other molecules) that generally are repeated at many places in the genome. This is one of the ways in which these genomic signals differ from the classical conception of a "gene" that expresses a particular trait. The use of repetitive elements to construct transcriptional regulatory circuits reflects two basic functional aspects of biochemical control:

- Most of the interactions between biomacromolecules tend to be relatively weak and need multiple synergistic attachments to produce stable, functional complexes [30, 61, 232].

- It is generally necessary to integrate the expression of different regions of the genome in a coordinated fashion to execute a particular phenotypic trait. This regulatory integration is often achieved by reusing the same binding sites at multiple locations [75, 201, 233, 234].

The dependence on weak molecular interactions and the synergistic nature of most molecular complexes provides dynamism and flexibility to the transcriptional regulatory circuitry because these complexes can be formed and taken apart easily [232].

From an evolutionary point of view, the main question to ask is how transcriptional regulatory circuits arise in the first place. How are similar binding sites amplified and distributed to multiple locations throughout the genome? How do higher-order circuit elements, enhancers, and more-complex *cis*-regulatory modules (CRMs) form and then disperse through the genome? These questions are distinct from those that evolutionists ask about the origins and diversification of coding sequences. We need to keep in mind that genomes contain many different kinds of information and that the entire cellular DNA has to evolve in a way that produces functional and adaptive expression systems. A little thought will make it clear how difficult it is to maintain the traditional idea that each individual component of these elaborate circuits evolves by making its own independent random walk through the enormous space of genome sequence possibilities. As you will see, there are alternative ways, based on established molecular processes, to think about the efficient evolution of genomic circuits based on rapid distribution of transcriptional regulatory sequence motifs.

Genome Compaction, Chromatin Formatting, and Epigenetic Regulation

For a few decades in the late 20th Century, it was possible to think that transcriptional regulatory circuits are sufficiently complex and sensitive to provide an adequate account of how cells regulate coding sequence access [68, 235]. However, two separate lines of thinking and experimentation taught us about additional layers of higher-order control on genome expression. Although originally independent, the two approaches converged at the molecular level in a surprising and satisfying way. This multilayered view of regulation enriches our ability to understand different aspects of genome function. It also turns out that it helps us account for the timing of evolutionary change in unexpected ways.

The first line of thought had its origins in the problems posed by multicellular development and cellular differentiation. How do cells become different from each other? How do tissues composed of specialized cell types form? What principles drive tissue formation and morphogenesis down well-defined paths during embryonic development? Without any detailed knowledge of molecular mechanisms, but well versed in developmental genetics, Conrad Waddington theorized about an "epigenetic landscape" that "canalized" genome function during development [236] [237–240]. Although described metaphorically in terms of surface grooves guiding marbles rolling down a hillside, this hypothetical concept has had lasting influence on researchers and has finally found a concrete molecular explanation.

The term *epigenetic* means "beyond (or added to) genetics." It refers to a mode of heredity independent of the basic DNA sequence or "genetic" constitution [9] [240, 241]. This idea is useful in understanding multicellular development. It describes how certain groups of cells—say, in a particular tissue or organ—can share inherited characteristics while retaining the same genome as cells with distinct inherited characteristics in a different tissue or organ [242–245].

Besides such theorizing about cell differentiation, several phenomena provided independent evidence for an additional mode of inheritance. Perhaps the most instructive of these cases is *genetic imprinting*, which in animals and flowering plants means that the expression of a genetic locus depends on whether it is inherited from the male or female parent [246] [247, 248]. Certain genetic loci are expressed only from the copy inherited from the father and other loci only when inherited from the mother [249]. In the mealy bug, where the term *imprinting* was first applied, the expression of a whole set of chromosomes inherited from the father is silenced in males but not in females [250]. Somehow, during the formation of the sperm and egg cells, different genetic loci or even whole chromosomes are marked, or *imprinted*, for silencing in the next generation.

Although the imprinting of a particular locus changes as it passes through male or female gametes, the underlying genetic information does not change. Because the imprint remains throughout multicellular development, the imprinted state is heritable through numerous mitotic cell divisions. In the mealy bug example, and in many vertebrate and plant examples, broader regions of the genome encompassing multiple genetic loci or whole chromosomes can be imprinted. Thus, epigenetic inheritance represents a high-level control (or set of controls) that can extend to entire haploid genomes [250, 251].

Epigenetic inheritance can extend beyond the development of a single individual to encompass several, or even many, generations. For example, the transgenerational epigenetic changes in maize plants called *paramutations* involve no alteration of DNA sequence but are stably transmitted through sexual reproduction for many generations [252]. Similar changes are found in animals [253–254]. In rodents, certain environmental stimuli, which include chemicals that disrupt endocrine signaling in sexual development, induce transgenerational changes in the offspring that are inherited by their descendants. Such heritable changes have been ascribed to alteration of epigenetic modifications [255–258]. Comparable environmentally induced transgenerational changes have also been documented in plants [259–261].

It has become evident that the epigenetic mode of inheritance exists in tandem with inheritance based exclusively on DNA sequences **[9]**. For a small but growing number of scientists, the *epigenome* (the constellation of all epigenetic modifications in the nucleus) constitutes a primary interface between environmental factors and the genome [262, 263]. As we turn our attention to the molecular nature of epigenetic modifications, you will see how this interface operates. Later, you will learn how epigenetic controls connect genome restructuring functions to organismal life histories.

The second line of research leading to our current understanding of epigenetics was experimentation in diverse fields on the relationship between DNA packaging into chromatin and replication and transcription. From early and mid-20th Century cytogenetics, it had been known that different chromosome regions stained distinctly and formed different types of chromatin (literally, "colored material"). Euchromatin ("true" chromatin) stained lightly and appeared to be the active region of the genome. Heterochromatin ("different" chromatin) stained darkly and was associated with silent regions of the genome (for example, the silenced paternal chromosomes in male mealy bugs) [264–266]. Genetic manipulations that placed an active genetic locus next to heterochromatic regions, such as centromeres, resulted in silencing of the previously active locus [267–269]. This "position effect" indicated that heterochromatic silencing could spread relatively long distances in the genome. Position effect phenomena further indicated that location in the genome was an important factor in controlling expression. Many experiments have verified the importance of "genome context" in the expression of individual genetic loci (such as [270]).

In the late 20th and early 21st Centuries, the correlation between chromatin structure and functional expression was placed on a defined molecular basis. The packaging of DNA within the eukaryotic cell depends at the most basic *nucleosome* level on winding the negatively charged double helix around positively charged proteins called *histones* [271, 272] [273, 274]. A special sequence code helps position the histones along the DNA to form regularly spaced nucleosomes [275–277]. There are two general differences between DNA and histones in euchromatin and heterochromatin:

- Heterochromatic DNA is more heavily modified by methyl groups attached to the cytosine (C) bases in its sequence.
- The histones in euchromatin and heterochromatin carry different chemical modifications (methyl, acetyl, and other chemical groups attached to particular amino acids in the histone "tails" that stick out from the nucleosomes) [278–280].

The histone modifications constitute what has come to be called a *histone code*; it allows molecular biologists to distinguish the chromatin state of associated DNA sequences [281–285]. By examining DNA methylation and histone modifications in active, silenced, and imprinted regions of the genome, a catalog is taking shape that allows us to identify the chromatin configurations associated with each epigenetic state of the genome. We already know from studies in yeast that the simple division into euchromatin and heterochromatin is too simple; chromatin configurations are specialized for different genomic functions [286–289].

Cells possess enzymes that either attach or remove methyl and acetyl groups from cytosines in DNA and exposed amino acid tails in histones [280, 290–292]. Thus, the formation or modification of chromatin structure is an active process with major consequences for the functional state of the underlying DNA [293, 294]. Such active *chromatin reformatting* is regulated by cell signaling circuits. It plays a major role in cell differentiation as cells become more specialized, silence large unused regions of their genomes by incorporation into silent heterochromatin, and open other regions encoding protein and RNA molecules needed for differentiated cell function [285, 295].

The ability of cells to target chromatin formatting within the genome is aided by the recently discovered role of micro- and other *noncoding* RNA molecules (ncRNAs) **[296]** [297–300]. Some of these ncRNAs form a complex with specialized RNA binding proteins linked to the enzymatic machinery for chromatin reformatting and specifically alter chromatin in the regions that have sequences complementary to the ncRNA [301–304]. You will see later how this RNA-targeted chromatin modifying/epigenetic regulation plays a critical role in the control of genome restructuring in response to episodes of cell stress or *genome shock*. Barbara McClintock used this phrase in speaking to explain a challenge or stress event that provoked a cell to activate the molecular systems that restructure genomes [207].

The *indexing* of the genome into extended chromatin domains that may encompass dozens of genetic loci and hundreds of thousands of base pairs is itself subject to additional formatting [251, 305]. We have learned about the existence of various classes of *insulator* sequences, which serve as boundaries between different types of chromatin [306, 307]. They also separate transcriptional formatting signals, such as promoters and enhancers [308, 309]. Some insulators nucleate *insulator bodies* that attach chromosomes to the nuclear envelope and thereby create a barrier to extension of chromatin domains [310, 311]. Other insulators work by directing RNA PolIII transcription of SINE or tRNA molecules and thus moving their chromosome site into one of many specialized *transcription factories* within the functionally compartmentalized nucleus [312–314]. Certain sequences format the initiation of silent chromatin domains. In general, these are repetitive DNA sequences recognized by ncRNAs [315–318]. These silent chromatin formatting sequences will become important when we discuss the impact that genome restructuring has on the expression of stored information.

Genome Formatting for Replication, Localization, and Transmission to Daughter Cells

To function effectively as a storage medium in proliferating cells, replicated DNA molecules must pass reliably to progeny cells at division. We have already discussed the role of checkpoint surveillance routines in ensuring the accuracy of this process in eukaryotic cells. As is the case for transcriptional regulation, DNA molecules must be formatted by the appropriate signals for a complete replication process, for proper localization during the cell cycle, and for accurate transmission at division. Several different functional types of formatting signals are involved in integrating these biochemical and biomechanical events into the cell cycle:

- **Sites for initiating DNA replication.** In prokaryotes these are called *origins of replication*, or *ori* sites [319–322]. For the basic genomic components, there is generally one *ori* per molecule. *Ori* sites tend to have a composite organization that includes multiple recognition sequences related to the control circuitry that ensures there is only one initiation event per cell division cycle. In eukaryotes, these initiation sites are called *autonomous replication sequences* (*ars*) and exist at multiple locations in each of the chromosomes. The *ars* sites are less defined in eukaryotes than *Ori* sites in prokaryotes, although recent work has begun to discern characteristic motifs [323]. They interact with an *origin of replication complex* (ORC) multiprotein apparatus that is connected to the cell cycle control circuitry so that each *ars* sequence can only initiate replication once per S phase of each cell cycle [324, 325].
- **Sites for completing DNA replication.** In prokaryotes, these assume a variety of forms. Many of the prokaryotic DNA molecules are circular in structure and contain special *terminus*

regions that have signals for biochemical processes that allow the replicated molecules to form completed duplexes and separate from each other if they have become interlocked or joined into a single double circle [322, 326, 327]. Some prokaryotes have linear DNA molecules with closed hairpin ends connecting the two strands (also found in some viruses). The hairpin ends contain signals that facilitate the "resolution" of daughter molecules by a special recombination event as soon as the hairpin has been replicated [328, 329].

In eukaryotes, the chromosomal DNA molecules are linear with open ends, which poses two problems for maintenance and replication:

    i.   The ends must be prevented from joining, in the same way that ends do for some forms of double-strand (DS) break repair.

    ii.   The replication apparatus can copy only one DNA strand completely, leaving the other strand incomplete.

These two problems are solved by constructing a special *telomere* (literally, "end body") structure at each extremity of the chromosomal DNA molecule. The telomere contains a number of different signals that facilitate the addition of extra DNA sequences to the end after replication (usually involving the enzyme telomerase but sometimes employing other mechanisms). They also format a special telomeric chromatin structure that protects against end-to-end joining by DNA repair functions [330] [331–337].

- **Sites for ensuring transmission at cell division.** In prokaryotes, these generally are labeled *partition* (*par*) sites. The *par* sites ensure separation of replicated DNA molecules and movement to the cell poles, powered by an actin-related motor protein, so that the two copies end up in separate cells after division in the middle of the cell [319, 338–340]. In eukaryotes, chromosome separation is formatted by the *centromere* (literally, "central body") or *cen* sequence. In most eukaryotes, centromeres are complex structures containing many tandemly repeated DNA sequence elements. These format a special centromeric heterochromatin structure that undergirds assembly of the *kinetochore* structure, which in turn attaches to the microtubules of the spindle apparatus. This ensures separation of sister replicated chromosome copies in mitosis [**148, 152**] [153, 341–343].

- **Sites governing subnuclear localization.** In addition to their basic functions in replication and distribution at cell division, we are also beginning to learn about the roles that centromeres, telomeres, and other DNA signals, such as insulators, play in localizing eukaryotic chromosomes within the nucleus at different stages of the cell cycle. We now recognize that the eukaryotic nucleus is a highly organized and subcompartmentalized organelle [**344**] [345–349]. Replication and transcription occur in separate *foci* or spots called replication and transcription *factories* [350–355]. Splicing and other RNA processing reactions are related to particular transcription factors, to visible intranuclear structures (or *granules*), and to the nuclear pores where the processed RNAs exit from the nucleus into other cell compartments [356–360]. Similarly, repair of DNA damage occurs in localized foci called "repair centers" [361–363].

We are at the very beginning of applying molecular cytogenetic methods to elucidate how important intranuclear position can be in genome functioning. However, several points are already clear and indicate that this will be a fruitful area of investigation:

- Classical cytogenetics has documented how important centromeres and telomeres are in localizing the chromosomes with respect to the spindle and nuclear envelope during the different stages of mitosis and meiosis [364–368].

- Application of molecular "chromosome painting" technologies, which allow the visualization of specific chromosomes, chromosome regions, or individual genetic loci, has established that chromosome positioning in interphase nuclei (during G1-S-G2 phases) is quite flexible but nonrandom [369–374].

- Interphase chromosome positioning patterns change in distinct tissues and cell types, and individual loci can be seen to alter their localization in response to specific stimuli. In certain cases, the movements or alignments of individual loci correlate with known molecular events, such as coordinated transcription [312, 358, 375, 376].

Distinct Classes of DNA in the Genome

In our discussion of genome formatting, we have distinguished between several functionally distinct classes of DNA and RNA sequence elements: coding sequences, noncoding sequences (which means that they do not encode protein but does not mean that they are devoid of information content), and formatting signals of various types. The bioinformatics specialists who analyze genome sequence data have different computational methods of classifying sequence elements based on accumulated experience in *annotating* (interpreting and assigning meaning to) and cataloging sequences. Their categories are revealing because they indicate that many genomes, such as our own, contain a very small proportion of what has traditionally been termed a "gene" (a unique protein coding sequence, either with or without its essential expression signals). Sequence analysis thus tells us that cells organize their genomes in different ways from those imagined by 20th Century geneticists. Functional studies, especially those that examine the spatial dynamics of genome action in living cells, complement sequence analysis in an effort to find a more realistic basis for formulations about the system architectures of these remarkable information storage organelles.

Table II.1 gives a 2005 listing of the genome compositions of a number of sequenced animal and plant genomes [62]. Table II.2 lists some of the various classes of DNA elements that have been annotated in genomes. The specific references for the information in these tables are available in the online versions of the tables. A major distinction is between unique and repetitive DNA sequences [**377**] [62, 75, 378–381]. One of the most striking discoveries of the Human Genome Sequencing Initiative was that the proportion of protein-coding DNA (exons) in our DNA is rather small (~1.5%) compared to other classes, such as dispersed repeats (~40%) and tandem repeats of various types (~25%) [382]. Even the highly reduced genomes of many bacteria contain significant proportions of repetitive DNA (over 10% in several species, such as *Neisseria meningitidis*, where the repeat content has been carefully analyzed) [383].

Table II.1 Different Classes of DNA in Selected Genomes (References Appear in the Online Version)

Species	Genome Size	Percentage of Repetitive DNA	Percentage of Coding Sequences
Animals			
Caenorhabditis elegans	100 MB	16.5	14
Caenorhabditis briggsae	104 MB	22.4	13
Drosophila melanogaster	175 MB	33.7 (female) ~57 (male)	<10
Ciona intestinalis	157 MB	35	9.5
Fugu rubripes	365 MB	15	9.5
Canis domesticus	2.4 GB	31	1.45
Mus musculus	2.5 GB	40	1.4
Homo sapiens	2.9 GB	>50	1.2
Plants			
Arabidopsis thaliana	125 to 157 MB	13 to 14	21
Oryza sativa (indica)	466 MB	42	11.8
Oryza sativa (Japonica)	420 MB	45	11.9
Zea mays	2.5 GB	77	1

Many of the repetitive elements contain signals that influence various functions wherever they exist within the genome. These functions include recombinational DNA repair, transcription initiation (promoters and enhancers in many dispersed repeats), modulation of transcriptional elongation (LINE elements), centromere formatting (tandem repeat arrays in many organisms), and attachment to the nuclear matrix (LINE elements). Two 2005 compilations of the functional consequences of repetitive DNA elements listed over 80 documented examples in which repetitive elements formatted one or more of all seven genome functions set out at the start of Part II [62, 384].

Table II.2   Different Classes of Annotated Repetitive Genome Components (Amplified from [62]) (References Appear in the Online Version)

Structural Class	Structural or Functional Characteristics
Oligonucleotide motif	4 to 50 bp. Protein binding or recognition sites.
Homopolymeric tract	Repeats of a single nucleotide $(N)_n$.
Variable nucleotide tandem repeats (VNTR)	Repeats of dinucleotides and longer sequences <100 bp that may vary in number in the tandem array: (NN...N)
Composite elements	Composed of two or more oligonucleotide motifs, sometimes with nonspecific spacer sequences. Examples include palindromic operators, promoters, enhancers and silencers, replication origins, and site-specific recombination sequences.
Tandem array microsatellites or simple sequence repeats (SSR)	Head-to-tail repeats of small sequence elements from two to six base pairs in length. Subject to frequent changes in repeat number and length. In genetic loci, expression levels tend to decrease with increased microsatellite length.
Tandem array satellites	Repeats of larger elements, typically 100 to 200 bp in length. Satellite arrays typically contain thousands of copies. Often found at centromeres.
Terminal inverted repeat (TIR) DNA transposons	DNA-based mobile genetic elements flanked by inverted terminal repeat sequences of ≤50 bp. May encode proteins needed for transposition. Vary in length from several hundred to several thousand base pairs.
Foldback (FB) DNA transposons	DNA transposons with extensive (many kb) inverted repeats at each end.
Rolling circle DNA transposons (helitrons)	DNA transposons that insert from a circular intermediate by rolling circle replication. Can generate tandem arrays.

Table II.2   Different Classes of Annotated Repetitive Genome Components
(Amplified from [62]) (References Appear in the Online Version)

Structural Class	Structural or Functional Characteristics
Long terminal repeat (LTR) retrotransposons	Retroviruses and nonviral mobile elements flanked by direct terminal repeats of several hundred base pairs. Insert at new locations following reverse transcription from an RNA copy into duplex DNA.
Long interspersed nucleotide element (LINE) retrotransposons	Mobile elements several kb in length with no terminal repeats. Encode proteins involved in retrotransposition from a PolII-transcribed RNA copy by target-primed reverse transcription.
Short interspersed nucleotide element (SINE) retrotransposons	Mobile elements a few hundred base pairs in length with no terminal repeats. Do not encode proteins (mobilized by LINE products from a PolIII-transcribed RNA copy).

One important feature of repetitive DNA elements is that they are better taxonomic markers than protein-coding sequences [381, 385, 386]. Organisms that share basically the same repertoire and structure of proteins may differ markedly in one or more categories of repetitive DNA elements. Examples include closely related species of *Drosophila* fruit flies, which have major differences in their tandem satellite repeats [387–389], and mammals, where the proteins are overwhelmingly similar but the families of interspersed SINE elements are strikingly different and strikingly abundant (at least tens of thousands of copies of each SINE family per haploid genome) [385, 390, 391]. This major taxonomic divergence between the specificities of repetitive and protein-coding sequences tells us that the genome's repetitive components are far more variable in evolution. Insofar as the repetitive elements in the genome format its functional architecture, the divergence also tells us that each taxonomic group may have acquired a distinct genome system architecture independently of changes in the encoded protein content.

From a systems view of genome organization (thinking of the genome as more functionally integrated than a collection of autonomous genetic units), it should be apparent that generic functions must be carried out at multiple locations. Because these functions all require protein-DNA interactions, it makes sense for the corresponding recognition sequences to be repeated throughout the genome [75] or, as in the case of centromeres and telomeres, to be localized at the right positions in different DNA molecules. Because the genome, like other complex organized systems, has a hierarchical structure, it is also evident that genomic elements comprising defined complexes of distinct formatting signals (such as transposons and retrotransposons) can operate as integrated microprocessors at many different places throughout the genome.

From an evolutionary perspective, a major question is, How do these repeated signals distribute themselves within and throughout the genome? If alterations in DNA molecules occur randomly, the problem of distributing generic formatting elements and larger complexes composed of several elements becomes extraordinarily difficult, if only in finding the time needed for so many chance events to occur. (For instance, note the presence of over three million dispersed repeat elements in the human genome [382].) However, if cells have ways to mobilize defined segments of the genome to novel locations, this problem can be solved through the operation of those genetic mobility systems. Therefore, let us turn our attention to what we know about the processes of *natural genetic engineering* (the phrase I use to denote the capabilities cells have to restructure their genomes).

The Molecular Mechanisms of Natural Genetic Engineering

Over the last 60 years, DNA has proven to be an extremely complex and malleable information storage medium. Virtually all cells possess the basic biochemical tools for modifying DNA: proteins that cut, unwind, polymerize, anneal, and splice DNA strands. The generic operations that living cells have been shown to carry out on their genomic molecules indicate that any rearrangement is possible as long as the product is compatible with the basic rules of DNA structure (see Table II.3).

Table II.3   Generic Cell Operations That Facilitate DNA Restructuring [30, 61, 392]

Polymerizing DNA strands complementary to a DNA or RNA template
Polymerizing DNA strands without a template (mutator polymerases and deoxyribonucleotide terminal transferases)
Cleaving phosphodiester bonds in the DNA strand backbone on one or both strands (exonuclease and endonuclease activities)
Resealing (ligating) cleaved phosphodiester bonds in novel combinations involving single-strand (SS) or double-strand (DS) DNA molecules
Cleaving and resealing phosphodiester bonds to connect RNA and DNA strands

In elaborating on these generic DNA operations, the vast majority of free-living prokaryotes and eukaryotes have evolved biochemical systems that permit them to mobilize and restructure their genomes in more specialized ways. Essential to any contemporary account of evolution is the inclusion of these well-documented genomic mobilization and restructuring (natural genetic engineering) operators (see Table II.4).

Table II.4   Natural Genetic Engineering Systems [6-8]

DNA import and export
Homologous recombination (HR)
Nonhomologous end-joining (NHEJ)
Site-specific reciprocal recombination (tyrosine and serine recombinases)
DNA transposons (replicative, cut-and-paste, helitrons)
Long terminal repeat (LTR) retroelements (retroviruses, retrotransposons)
Non-LTR retroelements (LINE, SINE, SVA retrotransposons)
Retrosplicing group II introns
Inteins and homing group I introns
Diversity-generating retroelements (DGR)

DNA Import and Export Systems

DNA import and export systems were in evidence at the origins of molecular biology. The first solid evidence for DNA as a carrier of hereditary information was the 1944 demonstration by Oswald Avery and his colleagues that DNA is the chemical nature of the *transforming principle* (first identified by Fred Griffiths in 1928

[393]) that transmits virulence characteristics from dead Pneumo-
cocci to nonvirulent but living bacteria of the same species [394]. We
now know in exquisite detail how DNA is imported into *S. pneumoniae* and many other bacterial species [395, 396]. Early studies on
bacterial genetics likewise depended on the active transport of DNA
from one cell to another by a process called *conjugation* [397], and
similar systems exist in *Archaea* [398].

In a remarkable example of how evolution makes repeated use of
complex inventions, basically the same molecular apparatus is used to
take up DNA from the extracellular environment and also to export it
from one bacterial cell to another [399–403]. The same system also
can transfer DNA from bacteria to other types of cells: fungal, plant,
and animal [404–407]. Transfers to yeast and animal cells are known
to occur only in special laboratory situations, but the transfer to plants
is part of a natural process that *Agrobacterium tumefaciens* (literally,
"tumor-causing soil bacterium") and other bacteria use to transfer
DNA into the nuclei of plant cells. There, the transferred DNA
induces the formation of tumors that nourish the infecting bacteria
[408–411]. You will see in Part III of this book what an important role
DNA import and export systems have played in the evolution of both
prokaryotic and eukaryotic organisms.

Homologous Recombination

Homologous recombination (HR) is an intricate multistep process
that fulfills a variety of functions in living cells **[118]** [120, 125]. It
repairs DS breaks in the DNA of cells that contain a second unbroken
copy of the damaged duplex [123, 412]. In bacteria, expression of HR
proteins is an integral aspect of the SOS DNA damage response. In
the production of haploid spores and gametes during meiosis in many
eukaryotes, HR carries out physical exchanges between chromosome
homologues and maintains proper chromosome alignment in the first
meiotic division [413–416]. These meiotic chromosome exchanges
(and equivalent exchanges following DNA uptake or transfer in
prokaryotes) form the molecular basis for studies of genetic linkage,
recombination, and the construction of genetic maps. HR exchange
events between repeat homologies in chromosomes also appear to be
a major source of chromosome restructuring in organisms (like our-
selves) rich in tandem and dispersed repeats (see Table II.1)

[417–421]. In many eukaryotic and prokaryotic microbes, HR is used to exchange genetic information between silent and expressed regions of a genome. These exchanges either turn on and off or diversify expression of coding sequences, often those whose products are recognized by the immune system. These regulatory and diversifying adaptations of HR will be discussed later.

Non-Homologous End-Joining

Non-homologous end-joining (NHEJ) repairs DS breaks in situations where there is no homologous copy for recombinational repair (such as during the G1 phase of a haploid yeast cell cycle). The NHEJ process involves several distinct components to recognize broken ends, process them for proper resealing, and then join them into a complete duplex [422–424].

Because processing of broken ends occurs, and because NHEJ may not always join broken ends from the same site, NHEJ is an inherently mutagenic process [425, 426]. When it joins ends from a single breakage event, NHEJ may create local sequence changes that involve insertion, deletion, or duplication up to a few dozen base pairs in length. When it joins ends from two or more breakage events, NHEJ may create chromosome rearrangements such as deletions, inversions, and translocations [427, 428]. The fact that broken ends from different chromosomes are mobilized to specialized subnuclear "repair centers" facilitates such interchromosomal NHEJ-mediated rearrangements [429].

Certain regularly occurring DNA restructuring events use the NHEJ apparatus to seal double-strand breaks. A prominent example includes the chromosome breakage and rejoining events that are basic to the operation of the mammalian adaptive immune system. We will discuss these immune system DNA rearrangements at length later in this part of the book.

Site-Specific Reciprocal Recombination

Site-specific reciprocal recombination is a process studied principally in prokaryotes, although one example occurs in yeast, and certain retrotransposons use this mechanism to integrate into eukaryotic chromosomes [430, 431, 481, 482]. This mechanism of reciprocal DS DNA exchange received its name because the *recombinases* that execute it act only at pairs of certain highly structured and specific recombination sites. At those paired sites, a recombinase protein executes the DS genetic exchange through an orchestrated sequence of cleavage and resealing events [432–435].

Intermediates in those DNA cleavage/resealing events are covalent linkages of DNA strands to an amino acid in the recombinase. Depending on the system, those DNA-recombinase linkages involve either a tyrosine or a serine residue [431, 436, 437]. Because the protein structures and nature of the DNA breakage and rejoining events are basically different for the tyrosine and serine families of recombinases, it appears that this very specialized mechanism of genetic exchange must have arisen twice, *independently*, in the course of prokaryotic evolution.

Bacteria use site-specific recombination for a wide variety of normal life-cycle functions that will be described in Table II.6. Site-specific recombination is also one of the major mechanisms used to insert imported DNA into the genome of a prokaryotic (and perhaps yeast) cell following intercellular transfer [438–440].

DNA Transposons

DNA transposons include the first mobile genetic elements discovered using cytogenetic methods by Barbara McClintock in the 1940s [4, 441, 442] and the bacterial elements discovered by molecular methods in the 1960s and 1970s **[5]** [6, 443]. Since then, DNA transposons have been found to be almost ubiquitous among both prokaryotes and eukaryotes.

DNA transposons are defined genetic segments that can move (transpose) from one (donor) site in the genome to another (target) site [226, 444–446]. The unifying feature of DNA transposons is that they transpose from donor to target sites purely at the level of DNA with no transcriptional intermediates. The segments are defined by special recognition sequences at both ends; most commonly, these sequences are *terminal inverted repeats* (TIRs) of about one to three dozen base pairs in length. DNA transpositions involve cleavage of one or both strands at the donor site, cleavage at the target site, and resealing of phosphodiester bonds between donor and target DNA. Typically, the donor strand cleavages are quite specific and thus preserve the transposon structure as it moves to new sites. The target site cleavages show a wide range of specificities, from a very small number of sites for some transposons to virtually ubiquitous throughout the genome for others. In at least one well-studied case, the choice of targets is subject to control by dedicated recognition proteins [447]. Target choice is not always by sequence recognition; some elements appear to insert preferentially into specific forms of DNA, such as replication forks or inverted repeat structures [448–450].

Evolution has shown remarkable inventiveness in the structures of DNA transposons and in the molecular mechanisms of DNA movement to new sites (Table II.2). The particular mechanism for a given element is governed largely by the *transposase* proteins that recognize transposon termini and initiate the transposition process [451–454]. The variables include cleavage of one or both strands at the donor site, excision or retention of the transposon at the donor site, and whether the transposon replicates during transposition. Some bacterial elements require a site-specific recombination event to complete replicative transposition. Others mobilize the transposon in a single-stranded form, using polymerization of the complementary strand to remove the transposing strand from the donor site and then to complete insertion into the target site [455]. It seems that any combination of biochemical steps is possible so long as the final product is a complete DNA double helix [226].

One important characteristic of DNA transposition is the ability of transposons to mobilize external DNA sequences. McClintock had noted this ability in her pioneering studies. This DNA mobilization ability was part of the evidence that made the initial molecular mechanism for replicative transposition in bacteria so convincing. The mechanism readily explained the structures of genetic fusions, deletions, and inversions previously shown to be mediated by various transposons [5] [456]. In addition, composite transposons mobilized sequences encoding various specialized biochemical functions by incorporating them into structures bounded on either end by transposons with their characteristic terminal sequences [457]. Experiments showed early on that any sequence can be synthetically engineered into a composite transposon by purely *in vivo* methods and thus rendered mobile [458–460].

As transposon research in *Drosophila* advanced, Bill Engels and his colleagues demonstrated how large-scale rearrangements resulted from the ability of a transposase to make DS breaks at defined positions in the genome (http://engels.genetics.wisc.edu/Pelements/index.html). NHEJ could join the broken ends in new ways to mediate the formation of deletions, duplications, inversions, and translocations [461–464]. From these and other results, it is safe to assert that DNA transposons and their associated DNA cleavage activities provide at least one well-established mechanism for major chromosome rearrangements observed in the course of evolution.

An interesting feature of DNA transposons in animals is that their activity is often restricted to germline cells, where it will be of greatest evolutionary impact. In the case of the best-studied *Drosophila* transposon, this restriction results from germline-specific splicing of the RNA that encodes the transposase protein [465, 466].

Long Terminal Repeat (LTR) Retroelements

Long terminal repeat (LTR) retroelements constitute one of the most abundant classes of mobile genetic elements in eukaryotes [467, 468]. The human genome has about 450,000 LTR retroelements (8% of the total DNA [382]). Retroelements are even more frequent in the genomes of other species. Retroviruses, such as HIV, belong to this class of mobile genetic element [469]. In addition, numerous LTR retrotransposons have lost the ability to infect new cells but still retain the ability to migrate intracellularly throughout the genome of the host cell.

The intensively studied life cycle of retroviruses and some retrotransposons has verified the process of alternating DNA and RNA genome forms first postulated by Howard Temin in the 1960s [**183, 470**] [471]. RNA genome production occurs by transcription from chromosomally integrated DNA *proviruses*. The provirus is the viral genome integrated into the genome of its host cell, typically in a latent form. Andre Lwoff (Monod's doctoral supervisor) developed the provirus concept as a result of his studies with bacterial viruses in the early 1950s [472] [**473**] [474]. The directly repeated LTR DNA structure characteristic of the provirus forms as a consequence of intricately organized nucleic acid rearrangements that occur during reverse transcription of the retroviral (or retrotransposon) RNA genome [469, 475–477]. The end product of this elaborate reverse transcription process is an LTR-flanked DS DNA copy that can insert into new genomic target sites. Intriguingly, the insertion process requires a retroelement-encoded *integrase* function that generally operates in the same fashion as transposase proteins [478–480]. This functional mix of a conserved integration mechanism coupled with a radically different form of genetic mobility, via an RNA intermediate, illustrates the ability of evolutionary processes to adapt and combine existing functionalities into new, more complex systems. A particular class of LTR retroelements utilizes functions related to tyrosine site-specific recombinases for chromosomal integration [476, 481, 482].

The phenotypic consequences of LTR retrotransposition are subject to intense study because many retroviruses are transmissible cancer agents. The first infectious cause of cancer to be identified (in 1910) was the Rous Sarcoma Virus (RSV) retroelement in chickens [483, 484]. These tumor virus investigations have revealed two basic aspects of retroviral genome engineering. The first is that many tumor viruses, such as RSV, can incorporate segments from cell RNAs into their genomes. RSV has picked up the spliced form of the cellular *c-src* transcript and added it as a *viral oncogene* (*v-src*) in addition to its complement of viral reproduction functions. Other tumor viruses have similarly acquired spliced forms of cellular RNAs in their genomes, but most of them have lost essential viral functions and thus have become defective retroviruses dependent on active "helper" viruses for their reproduction and spread [469, 485]. We do not yet fully understand how incorporation of processed cell RNAs into retroviral genomes occurs, but nonviral RNAs are known to be present in retrovirus particles, and their incorporation into the provirus DNA has been reproduced in laboratory experiments [486].

Some LTR tumor retroviruses do not have viral oncogenes. Instead, they induce tumor cell formation by a process of insertional mutagenesis and modification of the expression of normal cell-cycle regulatory proteins [487–489]. Almost always, the insertion acts to increase or otherwise misregulate expression. Increased expression occurs because the retrovirus LTR structure contains a dense complex of transcriptional signals—promoters, enhancers, and terminators—and the LTR enhancers stimulate transcription from the cellular oncogene (*c-onc*) promoter [488, 490]. The role of retrotransposon LTRs as mobile enhancers has also manifested itself in yeast genetics, where retrotransposon insertion was observed three decades ago to be one of the most frequent genetic changes leading to increased expression of individual genetic loci [491, 492].

In *Drosophila*, LTR retroviruses and retroelements are major agents of morphogenetic mutations, such as changes in body color, the formation of altered or extra wings, or homeotic transformations of one appendage into another. Although some of these mutations may reflect enhancer activity, the most frequent agent of so-called "spontaneous" mutations with visible phenotypes in laboratory strains is the *gypsy* retrovirus [493–498]. Intriguingly, the *gypsy* genome contains a strong insulator element, and many of its phenotypic effects can be ascribed to the insulator separating normally interactive promoter and enhancer sequences [310, 499–502].

Non-LTR Retroelements

Non-LTR retroelements are the most abundant components of the human genome (over 2.3 million copies constituting about 34% of the DNA [382]). These elements use a mechanism of reverse transcription that does not generate an LTR repeat structure at each end. Canonical non-LTR retrotransposons come in two basic forms identified initially by bioinformatics analysis: LINE (long interspersed nucleotide element) and SINE (short interspersed nucleotide element) (see Table II.2). The LINE elements are widespread in plant and animal genomes, and a number of these retrotransposons were initially identified by genetic studies in *Drosophila* [503–508].

LINE retrotransposons typically are a few thousand base pairs in length, contain internal transcription signals for RNA polymerase II, and encode two proteins involved in reverse transcription (see Table II.2). The SINE elements are shorter (typically 100 to 300 base pairs in length), related in sequence to stable cell RNAs (tRNAs, short rRNAs, and protein export particle 7S RNA), and contain an internal RNA polymerase III promoter (see Table II.2). SINE elements do not contain coding sequences and depend on LINE element-encoded proteins for their reverse transcription and insertion into the genome.

The LINE/SINE reverse transcription and genome insertion process is highly nonspecific. All that appears to be obligatory for recognition by the requisite LINE-encoded proteins is the A-rich terminus of the inserting RNA, not the upstream sequences with specific classes of genomic information. This lack of specificity allows the same molecular apparatus to reverse-transcribe and insert a wide variety of other cell RNA molecules into the genome. Such indiscriminate LINE-mediated integration explains the origins of DNA sequences that correspond to spliced or otherwise processed RNA molecules throughout different genomes [222, 509–512].

Surveying the LINE repeats in the human genome has revealed that at least one third of all copies are attached to downstream sequences from the donor site [513]. What has happened, apparently, is that transcription did not terminate at the normal termination site for the LINE element but continued into the adjacent regions of the genome, where it terminated later. Such "read-through" transcription produced an extended LINE transcript that was later reverse-transcribed and integrated into a new target site. This so-called *retrotransduction* process can mobilize relatively short sequences through the genome and can be observed to occur in laboratory experiments [514–516]. In addition to the ability of LTR retroelements to incorporate segments of cellular RNA, retrotransduction demonstrates the ability of LINE-mediated events to mobilize short genomic segments to new locations. It seems that a division of genome restructuring labor exists between DNA transposons and retrotransposons. DNA elements mediate large-scale rearrangements (up to 4 MB transpositions have been documented [517]), while retroelements mediate small-scale mobilizations.

In addition to LINEs and SINEs, genomes contain composite SINE elements composed of segments from various origins. The best-characterized composite SINE occurs about 3,000 times in the human genome and has been given the label SVA (satellite sequence repeat $(CCCTCT)_n$ combined with VNTRs and Alu-like antisense segments) [518–521]. The SVA elements appear to have arisen and remained active in recent human history and to have been involved in genetic changes that separate humans from closely related primates, such as chimpanzees [520]. SVAs have the fascinating natural genetic engineering property of retrotransducing upstream (rather than downstream) genome sequences [522, 523]. This indicates that SVA transcripts originate at upstream genome promoters and terminate at the end of the SVA element. Naturally, the SVA upstream retrotransduction capacity adds to the genomic restructuring repertoire available for evolutionary innovation [524, 525].

Retrosplicing Group II Introns

Retrosplicing group II introns were initially discovered in the dairy industry organism *Lactococcus lactis*[2] [526–528]. Group II introns are common in other prokaryotes and in mitochondria and chloroplasts of eukaryotes and have been identified in an animal genome [528–530]. Group II introns are self-splicing catalytic RNA molecules (or *ribozymes*) **[531]** [532–534]. The splicing reaction is reversible at the RNA level (the excised intron can reinsert itself into the spliced transcript), and the same reaction can occur on the corresponding plus strand of the coding DNA [533, 535, 536]. Reverse-splicing into DNA creates an RNA loop that can be converted into a fully integrated DS DNA copy of the intron by any one of several alternative processes, including LINE-like reverse transcription or homologous recombination [537–540].

Reverse splicing into DNA is stimulated by an intron-encoded protein that has reverse transcriptase, RNA maturase, and DNA endonuclease domains [541, 542]. The specificity of the insertion process is determined by base-pairing between complementary nucleotides in the intron and the target (intron-free) insertion site [543]. However, specificity is not absolute, and the retrosplicing introns can insert into ectopic sites [544, 545]. Active sites of DNA replication and transcriptional terminators are preferential targets for the reverse-splicing reaction [546, 547]. Moreover, the targeting nucleotides within the intron can be engineered to direct insertion into other genomic sites [548–552]. In thermophilic prokaryotes, an active process of type II intron reverse splicing with relaxed insertion

---

2 A bacterium of such economic significance that it has been nominated as the State Microbe of Wisconsin.

specificity has recently been described [553]. Every insertion into a new site creates a novel intron. It is worth noting that the potential of mechanistically similar reverse splicing reactions for distributing dispersed DNA repeats throughout a genome has yet to be fully explored.

Inteins

Inteins are self-splicing protein segments that excise themselves from larger protein molecules and are found in all types of organisms [554–556]. They are natural genetic engineering agents because many acquire endonuclease activity after excision and can cleave the genome so that a copy of the intein coding sequence can be inserted at the site of cleavage [557]. Generally, the cleavage occurs uniquely at an empty target site encoding the spliced protein, and intein DNA is inserted by homologous replacement from a filled site. This kind of "homing reaction" spreads the intein in organisms that have both empty and intein-coding versions of the same coding sequence. The filled site is no longer sensitive to the endonuclease activity, because the insertion disrupts the cleavage sequence normally recognized by the excised protein.

The intein-encoded endonucleases are included in the group of highly specific *homing endonucleases* that are also encoded by certain self-splicing group I introns and by a subgroup of LINE elements, which insert into unique spacer sites in ribosomal RNA coding regions (rDNA) [558, 559]. Both group I introns and rDNA spacer LINE elements use the same strategy as inteins to spread through genomes. When a homing endonuclease-coding segment occupies all empty sites in the genome, it loses functional targets and can degenerate. This means that a number of endonuclease-negative group I introns may be degenerate forms that originally evolved as endonuclease-coding elements.

In prokaryotes, inteins are found inserted into sequences encoding key genome maintenance proteins, such as DNA polymerase, other replication activities, and RecA. Fungi display a variety of different intein insertions in RNA polymerase subunits, translation initiation factors, and RNA splicing functions [554, 560, 561]. How and why these intein insertions occurred in the first place remains a mystery, but their presence in such key genome maintenance and expression proteins suggests some yet-to-be-discovered regulatory role.

Diversity-Generating Retroelements

Diversity-generating retroelements (DGRs) were recently discovered in a virus that infects the bronchitis pathogenic bacterium *Bordetella bronchisepticum* [562, 563]. The virus utilizes a reverse transcriptase-based mechanism to diversify the sequence of its primary cell attachment protein and thereby adapts to infecting different strains of bacteria. In addition to reverse transcriptase, the DGR system involves additional essential components. One of these is an accessory coding sequence that ends in a *template repeat* (TR) segment, which is quite similar to the *variable repeat* (VR) region in the coding sequence targeted for change. The DGR process specifically modifies adenine (A) residues in the TR region and transmutes them to other nucleotides that end up in the VR region, while the TR itself remains unchanged. Apparently, the TR region is transcribed, and then the corresponding RNA is reverse-transcribed in a LINE-like reaction that replaces the resident VR region with a new sequence [564].

Searching bacterial genomes has revealed similar DGR structures in at least 40 diverse species, ranging from pathogenic spirochetes such as *Treponema* (the syphilis pathogen) to human probiotic commensals such as *Bifidobacterium* to photosynthetic green sulfur bacteria (*Chlorobium, Prosthecochloris*) and cyanobacteria (*Trichodesmium, Nostoc*). Many of these bioinformatically identified DGR systems are associated with plasmid or host protein coding sequences rather than viral sequences [565]. Unfortunately, none of these other systems has yet been the subject of detailed analysis.

Natural Genetic Engineering as Part of the Normal Life Cycle

A major assertion of many traditional thinkers about evolution and mutation is that living cells cannot make specific, adaptive use of their natural genetic engineering capacities. They make this assertion to protect their view of evolution as the product of random, undirected genome changes. But their position is philosophical, not scientific, nor is it based on empirical observations. This section demonstrates that in a large number of well-documented cases, natural genetic engineering capabilities have been utilized as part of the normal organism life cycle. In many of these cases, utilization involves the integration of different natural genetic engineering processes into a highly targeted and well-regulated series of changes with a clear adaptive benefit. The operation of a tightly regulated sequence of natural genetic engineering events in the adaptive immune system is probably the most elaborate example we have of purposeful genome manipulations. If, as you will see in this section, cells can integrate processes of genome restructuring to serve adaptive needs in normal life cycles, there is no scientific basis on which to argue that cells cannot also use those same functional capacities to produce significant evolutionary novelties.

Site-Specific Recombination, Homologous Recombination, and Diversity-Generating Retroelements in Phase and Antigenic Variation

Many microorganisms need to change the repertoire of surface proteins they produce (see Table II.5) [566]. These proteins determine where cells attach to surfaces or to other cells, and they also constitute a major target of immune system defenses by host organisms. Conditions repeatedly change during a microbe's life cycle: from one host (such as an insect vector) to another (mammal), from life as an intestinal commensal to life in the external environment, or as the host immune response recognizes the microbe's surface antigens.

To adapt to these changes, both prokaryotic and eukaryotic microbes (such as Gonococci, spirochetes, or trypanosomes) use a variety of natural genetic engineering functions to turn on and off the synthesis of specific proteins (*phase variation*) or to change the proteins' structures (*antigenic variation*):

- Proteins are turned on and off by site-specific recombination between oppositely oriented recombination sites. This process inverts either promoters or coding sequences so that they align (turn on expression) or face in opposite directions (turn off expression) [567–571].

- Homologous recombination can exchange different protein-coding information in cassettes flanked by homologous sequences. The cassettes can be located at an *expression site*, where the coding information is transcribed and translated, or they can be warehoused in silent form at a storage location in the genome. Recombination from a silent cassette into the expression site modifies the expressed information. If a silent cassette contains nonfunctional coding information, the recombination event turns off expression. If a silent cassette contains alternative coding information, the recombination event produces a novel protein structure and antigenic variation. In the spirochete *Borrelia* (Lyme disease, relapsing fever) and in trypanosome worms (sleeping sickness, Chagas' disease), these antigenic variations can proceed through hundreds of sequential changes to keep the infecting population one step ahead of the immune system [572–576].

- Site-specific recombination can also produce antigen variation when oppositely oriented recombination sites are embedded within protein coding sequences. A recombination event reorients the DNA between the sites and replaces the original coding strand with its nonidentical complementary strand, thereby altering protein structure. There are numerous examples of such protein-diversifying *shufflons* in the prokaryotic genome sequence database [577–580]. Some shufflons contain as many as seven different recombination sites. Inversions between these multiple sites can generate dozens of different protein coding sequences. An intriguing mystery is how protein coding sequences evolve to contain two or more site-specific recombination signals.

- The role of diversity-generating retroelements (DGRs) in protein modifications has already been described. It remains to be determined how many of these reverse transcriptase-based change systems are utilized for antigenic variation and how many for functional diversification of the target proteins.

Table II.5   Control of Bacterial Protein Synthesis (Phase Variation) and Modification of Protein Structure (Antigenic Variation) by Natural Genetic Engineering (Expanded from [566]) (References Appear in the Online Version)

Phase Variation by Site-Specific Recombination

Escherichia coli (intestinal flora), *Moraxella bovis* (bovine pathogen), and *Moraxella lacunata* (human pathogen) (all Gammaproteobacteria): type I pilus synthesis

Mycoplasma pulmonis (Mollicute mouse pathogen): DNA restriction and modification

Pseudomonas fluorescens (Gammaproteobacteria plant pathogen): root colonization function(s)

Salmonella enterica serovar Typhimurium (Gammaproteobacteria mouse pathogen): flagellar synthesis

Clostridium difficile (Firmicute human intestinal pathogen): major cell wall protein

Campylobacter fetus (Epsilonprotebacteria human pathogen): surface layer protein

Developmental Activation of Expression by Site-Specific Excision from Interrupted Coding Sequences in Terminally Differentiated Cells

Bacillus subtilis (Firmicute): SigK expression in spore mother cell

Anabaena (*Nostoc*, cyanobacterium): NifD and FdxN expression in specialized nitrogen-fixing heterocysts

Phase Variation by Transposon Insertion and Excision

Acidithiobacillus ferrooxidans (Gammaproteobacteria soil bacteria): iron oxidation and swarming functions

Pseudoalteromonas atlanticus (Gammaproteobacteria marine biofilm organism): extracellular polysaccharide synthesis (IS492 insertion and excision)

Neisseria meningitides (Betaproteobacteria human pathogen), *Citrobacter freundii* (Gammaproteobacteria opportunistic human pathogen): capsule synthesis

Legionella pneumophila (Gammaproteobacteria human respiratory pathogen): lipopolysaccharide synthesis

Shigella flexneri (Gammaproteobacterial human dysentery pathogen): cell surface markers

Staphylococcus aureus, Staphylococcus epidermidis (Firmicute human pathogens): extracellular polysaccharide synthesis and biofilm formation

Xanthomonas oryzae (Gammaproteobacteria plant pathogen): extracellular polysaccharide and virulence

Phase Variation by Cassette-Based Recombination

Geobacillus stearothermophilus (Firmicute soil bacterium): S-Layer proteins

Table II.5 Control of Bacterial Protein Synthesis (Phase Variation) and Modification of Protein Structure (Antigenic Variation) by Natural Genetic Engineering (Expanded from [566]) (References Appear in the Online Version)

Antigenic Variation by Cassette-Based Recombination

Borrelia burgdorferi (Lyme disease Spirochaete), *Borrelia hermsi* (relapsing fever Spirochaete): surface lipoproteins

Helicobacter pylori (Epsilonproteobacteria gastric pathogen), outer membrane proteins

Mycoplasma synoviae (Mollicute avian pathogen), *Mycoplasma genitalium* (Mollicute human pathogen): surface lipoproteins

Neisseria gonorrhoea (Betaproteobacteria human pathogen): opacity proteins (Opa) and type IV pili

Treponema pallidum (syphilis Spirochaete): major surface antigen

Anaplasma marginale (intracellular Rickettsial pathogen): immunodominant outer membrane protein

Antigenic Variation by Site-Specific Recombination

Bacteroides fragilis (intestinal microflora), polysaccharides

Campylobacter fetus (Epsilonprotebacteria opportunistic human pathogen), surface proteins

Dichelobacter nodosus (Gammaproteobacteria sheep pathogen), outer membrane proteins

Mycoplasma pulmonis (Mollicute mouse pathogen), *Mycoplasma penetrans* (Mollicute opportunistic human pathogen), *Mycoplasma bovis* (Mollicute cattle pathogen): surface lipoproteins

Bacteriophage Mu G tail protein

Plasmid R64 conjugative pilus shufflon

Other shufflons in genome sequences

Antigenic Variation by Diversity-Generating Retroelements (DGRs)

Bordetella bronchiseptica (bronchitis pathogen): bacteriophage tail fiber

Other DGRs in the genomes of a marine *Vibrio* virus and also in the chromosomes of a commensal *probiotic Bifidobacterium*, the dental spirochete *Treponema denticola*, and five different cyanobacteria

Site-Specific Recombination in Temperate Phage Life Cycles, Chromosome and Plasmid Replication, and Terminal Differentiation of Bacterial Cells (Table II.6)

The first site-specific recombination system to be identified and analyzed belongs to the bacterial virus (*bacteriophage*) λ, the one used by Weigle in his mutagenesis studies. λ is designated a *temperate bacteriophage* because it has a complex life cycle that includes the dormant *prophage* state described by Lwoff [473]. Molecular genetic analysis showed that the λ prophage genome exists integrated into the *E. coli* chromosome [581]. The temperate bacteriophage genome thus exists in one of three states: inactive inside the virus particle, actively reproducing inside virus-producing cells, or inactive as a repressed *provirus* integrated in the bacterial genome. Conversion between the actively reproducing and prophage states involves insertion into and excision out of the *E. coli* chromosome mediated by site-specific recombination [582, 583]. For λ and many other temperate bacteriophages, integration and excision occur in a tightly regulated manner linked to coordinated controls on expression of viral reproduction functions [584, 585].

During the bacterial cell cycle, site-specific recombination plays an active role in the replication and distribution of bacterial DNA molecules, chromosomes, and plasmids (collectively called *replicons*, or replicating entities). Bacterial chromosomes and many plasmids that replicate as circular DNA molecules contain the coding sequence for a tyrosine or serine recombinase and its cognate recombination site. It often happens that recombinational exchange during replication of such molecules produces a double circle rather than two separate circles. When this happens, the recombination site is duplicated within the same molecule. This duplication creates the conditions for site-specific recombination to separate the two copies into distinct molecules, which can then be transmitted independently to daughter cells [586].

A third role for site-specific recombination in the normal life cycle occurs in bacteria that undergo terminal differentiations. Two examples have been studied in detail: the "mother" cell that helps form a dormant spore during *Bacillus subtilis* sporulation [587], and the nitrogen-fixing *heterocyst* that forms as a result of nitrogen starvation of photosynthetic cyanobacteria [588]. In both cases, the differentiated cell does not divide further but contributes essential functions to companion cells that have to survive under starvation conditions and then proliferate. In both cases, the coding sequences for functions essential to the differentiated state are interrupted by DNA segments that are removed by site-specific recombination during the differentiation process [589, 590]. The excised DNA does not have to be replaced because the terminally differentiated cells leave no progeny, and the excision does not occur in the companion cells that will go on to future divisions.

Table II.6 Applications of Site-Specific Recombination to Different Functions in Bacterial Cells [591] (References Appear in the Online Version)

Integrate infecting viral genomes, which can later be excised by site-specific recombination.

Integrate and excise single-protein coding cassettes for antibiotic resistance and other cell properties into expression structures called *integrons* or (in the case of very large structures encoding diverse proteins) *superintegrons*.

Integrate horizontally transferred DNA segments (genomic islands).

Separate intermediate structures in the movement of DNA transposons.

Resolve tandemly repeated chromosomes and smaller replicons into two separate molecules for proper distribution to daughter cells.

Resolve replicated telomeres on prokaryotic linear chromosomes.

Invert DNA segments to regulate transcription (see Table II.5).

Invert DNA segments to alter protein coding sequences (see Table II.5).

Excise DNA introns to permit the expression of specialized functions in terminally differentiated bacterial cells (see Table II.5).

Mating Type Switches in Yeast

Mating type switches in yeast are among the most fascinating and well-studied examples of natural genetic engineering in a normal eukaryotic life cycle. Both budding (brewer's) yeast *Saccharomyces cervisiae* and fission yeast *Schizosaccharomyces pombe* (used to make beer in Africa) can change the sex of a haploid cell by transferring genetic information from silent mating type cassettes to the expressed mating type locus [592–594]. The sex change operation permits the progeny of a single haploid spore to generate cells of both mating types, which can then mate and produce diploid progeny (see

Appendix I.2 online). Diploid cells have at least one advantage over haploids. Even in the G1 phase of the cell cycle, diploids have two copies of the genome and can use one copy as a template to repair damage to the other in an error-free recombination process.

Both yeast species share several features of mating type determination and switching:

- The difference between expressed and silent mating type information is maintained by placing the silent cassettes in specialized chromatin domains.
- The transfer of information to the expressed mating type locus involves general homologous recombination functions.
- The information transfer process is initiated and targeted specifically to the expressed mating-type locus by a highly regulated DS break at a specific site.
- Information transfer virtually always goes from the silent cassette holding information for the opposite sex unidirectionally to the mating type locus.

In both *S. cerevisiae* and *S. pombe*, a general homologous recombination process has been domesticated for use as a targeted unidirectional transfer of information. What is intriguing is that there appear to be several mechanistic differences in the recombination process. One notable difference observed among yeasts is that *S. cerevisiae* makes its initiating DS break at the mating type locus by a homing endonuclease type of protein [595, 596], while the closely related species *Kluyveromyces lactis* uses a modified transposase protein [597]. The differences between the switching process in these three species indicate that the integration of DS break functions, directional control of information transfer, and homologous exchange mechanisms arose independently more than once in yeast evolution [598].

Macronuclear Development in Ciliate Protozoa

Macronuclear development in ciliate protozoa (such as *Paramecium*) is probably the most massive and amazing rapid genome restructuring known [599–603]. Within a single cell generation, a whole new genome governing mitotic cell proliferation is constructed out of a germline genome organized in a completely different way. The cells of ciliate protozoa have two nuclei. A small nucleus, called the *micronucleus*, is transcriptionally silent but carries the germline chromosomes. A large nucleus, called the *macronucleus*, is transcriptionally active and produces the functional RNAs for vegetative cell growth by mitosis. In effect, ciliate protozoa have separate germline and somatic nuclei within the same cell.

When a culture of ciliate protozoa is deprived of nutrients, the cells respond by undergoing the following sequence of events:

1. Meiotic division of the germline micronucleus to produce haploid gamete nuclei

2. Mating and exchange of gamete nuclei by two postmeiotic cells

3. Gamete fusion in each mated cell to produce a diploid zygote nucleus

4. Division of the zygote nucleus

5. Degeneration of the old macronucleus

6. Development of one of the two sibling zygote nuclei into a new macronucleus

Macronucleus development from a zygote nucleus involves a remarkable series of cellular and genomic events:

1. The germline chromosomes undergo a process of endoreplication without cell division. This produces a series of side-by-side multicopy *polytene* (literally, "many bands") chromosomes that look very much like those from *Drosophila* salivary glands.

2. Membrane vesicles surround segments of the polytene chromosomes while they undergo a process of fragmentation into thousands of separate DNA molecules (the size and number of the fragments vary from species to species).

3. The chromosomal DNA is progressively fragmented. The DNA that is removed comprises the *internal eliminated sequences* (IES), and the DNA that is retained includes the *macronuclear destined sequences* (MDS). We now know that the MDS component is determined by the RNA molecules found in the old macronucleus, but the exact mechanism of RNA guidance is not yet clear [604, 605]. The discarded IES

component typically is >90% of the germline genome. The remaining MDS component consists of thousands of small minichromosomes, each in many copies.

4. Each macronuclear minichromosome appears to encode a single protein. Some MDS/minichromosome segments are already present in final form in the germline chromosomes and only need to be excised (cut out) by nuclease activity. However, many macronuclear minichromosomes have to be assembled by joining multiple MDS segments. In quite a few cases, including the sequences that encode DNA polymerase and other essential proteins, the component MDS segments are present in a "scrambled" arrangement in the germline chromosomes (in a different order and orientation than they will end up in the macronuclear coding sequence). These scrambled MDS segments can number in the dozens for a single coding sequence, and they have to be properly excised, unscrambled, aligned, and joined for the macronucleus to function. The details of this fascinating unscrambling process remain to be defined, but the guidance of RNA from the old macronucleus apparently plays a key role in accurately aligning the pieces [604–610].

5. As soon as all the final multicopy macronuclear minichromosomes have been produced, the telomerase enzyme adds telomere repeats to each end so that the minichromosomes can replicate completely. The reason telomerase was first isolated in 1985 by Carol Greider and Elizabeth Blackburn from *Tetrahymena*, a ciliate, was that the macronucleus has an abundance of the enzyme necessary to maintain the many thousands of telomeres it contains [611] **[330]**. (In contrast, the nucleus of a human cell has only 92 telomeres, two for each of our 46 chromosomes.) Telomerase is a specialized form of reverse transcriptase that polymerizes telomere repeats from an RNA template [612].

6. The multicopy macronuclear chromosomes have no centromeres, but they can maintain themselves for hundreds of cell divisions in the laboratory. Some natural ciliate isolates

have been found that lack a micronucleus. This demonstrates both that the macronucleus alone is essential for vegetative growth and that macronuclear minichromosomes are stably inherited in the wild. How they are stably maintained and distributed at macronucleus division remains a mystery.

Whenever anyone challenges the ability of living cells to carry out biologically functional, rapid, and massive genome restructuring, they should learn about the ciliates. Every round of starvation and sexual exchange demonstrates that these organisms reliably use thousands of DNA cleavage and resealing events to convert a meiotically functional genome into a structurally different somatic genome in a matter of several hours. Massive and rapid genome restructuring is occurring around us all the time.

Does LINE Element Retrotransposition Play a Role in Somatic Differentiation of the Mammalian Nervous System?

Another possible role of natural genetic engineering in the normal life cycle of multicellular organisms is to generate somatic diversity within tissues [613]. A number of investigators are postulating that induced genome change by LINE-1 retrotransposition may be linked to the establishment of neuronal identity in higher nervous system development [614–621]. The intriguing hypotheses about nervous system development are the result of recent and expanding observations that LINE-1 retrotransposition is far more frequent in neural and brain tissue cells than in other cell types. Moreover, the higher levels of LINE-1 mobility apparently respond to genomic control factors, such as the Wnt and MeCP2 regulatory proteins, that have been implicated in neurological phenotypes such as Rett syndrome (developmental regression) and autism [622–625]. It also has been suggested that somatic tissue LINE retrotransposition may participate in aging [626] and cancer [627].

The Mammalian Adaptive Immune System

The mammalian adaptive immune system probably has the most sophisticated functional DNA rearrangement apparatus known to biology. One of the reasons we are so familiar with these DNA rearrangements, of course, is that adaptive immunity is of enormous importance in disease and disease prevention.

A key point to recognize is that the adaptive immune system uses an evolutionary process to solve a very difficult problem: how to generate a series of properly structured proteins that can recognize and bind to a virtually infinite and largely unpredictable range of invaders. The germline genome has finite DNA and can encode only a limited range of proteins. To make the requisite number of distinct antigen recognition proteins, the lymphocytes have to evolve rapidly. In other words, they have to generate tremendous antigen-binding variability while maintaining proper protein structure for interaction with other components of the immune response. Lymphocytes achieve their evolutionary goal through the use of targeted but flexible natural genetic engineering steps.

In discussing this highly evolved rapid evolution system [607, 628–630], we will focus on the part of it that generates and then modifies the coding sequences for antibody molecules produced by the B lymphocytes. We will neglect other antigen recognition components, such as the T cell receptors. There have been numerous accounts of the DNA restructuring and targeted mutagenesis processes involved in antibody production during the immune response. If you're particularly interested, many of the details (and the specific literature references) are presented online in Appendix II.2. Here we'll summarize the details and see what general lessons about evolution we can learn from the production of antibodies in adaptive immunity. In other words, let's see how evolution itself evolved a rapid evolution process. The articles by Edelman, Leder, Tonegawa, and Nossal in the Suggested Readings cover some of the same ground [631–634].

The evolving B lymphocytes use a combinatorial process based on DNA breakage and rejoining. These molecular events put together complete antibody coding sequences by choosing from among hundreds of cassettes that encode alternative versions of "variable regions" in the two protein chains of the antibody molecule.

(Each antibody contains two "light" chains and two "heavy" chains.) About 48 million possible cassette combinations are possible. Overall antibody protein structure and functionality are maintained because DNA breakage around the cassettes occurs at specific signals. We now know that the sequence-specific DNA cleavage reactions evolved from DNA transposons and that the *RAG* proteins that carry out these cleavages do, in fact, operate in yeast cells as transposases [635].

Additional antibody diversity comes from molecular flexibility in the cassette-joining process (*junctional diversity*). The same two cassettes can combine to produce a dozen or more distinct coding sequences. Still further diversity arises when untemplated DNA sequences are inserted between the cassettes that encode the variable region of the larger "heavy" immunoglobulin chain. Altogether, the combined sources of variability produce well over a million million (10^{12}) different antibody specificities.

Newly made antibodies sit on the surface of the producer B lymphocyte, where they are connected to the circuitry that controls cell proliferation. When a surface-bound antibody binds an antigen, the producing B cell is activated to proliferate. This proliferation selectively amplifies clones of cells that produce still more antibody having the right specificity for a particular invader.

Antigen-dependent B cell stimulation doesn't just trigger proliferation. It also alters the behavior of the producing B lymphocytes. So-called *activated B cells* migrate to new locations in the lymphatic tissues and undergo two further evolutionary steps. One step, called *somatic hypermutation*, is the targeted mutagenesis of the DNA regions that encode the antigen-binding "variable" regions of the two antibody proteins. Following somatic hypermutation, some mutagenized antibodies bind the antibody with greater affinity and will be even more strongly stimulated to proliferate. Thus, targeted mutagenesis fine-tunes and improves the repertoire of antigen recognition proteins.

The second step in activated B cell antibody evolution involves a change in the "constant region" of the "heavy" chain. The constant region is the part of the antibody molecule that does not undergo somatic hypermutation and, independently of antigen recognition specificity, interacts with cell structures and with other components of the immune system. These interactions define the antibody *isotype*, which characterizes where it is located (such as in the bloodstream or on mucosal surfaces) and how it binds to other parts of the immune system (such as killer cells or the complement system).

Constant region (*isotype*) switching is accomplished by DNA breakage and rejoining events that are quite different from those used in combining the variable region cassettes. Isotype switching occurs at specific *switch* regions and is targeted by *lymphokine* signaling proteins secreted by other immune system cells. The various lymphokine signals tell the activated B cell where to transcribe upstream of the different switch regions preceding distinct constant region exons. This transcription leads to cleavage of the transcribed DNA. By choosing a particular constant region exon for rearrangement, the lymphokine signaling determines the antibody isotype that is produced after NHEJ joining of the broken ends.

These amazing examples of functionally dedicated and targeted DNA changes offer some key lessons about natural evolution:

- The initial genomic inventions only need to work at base-level effectiveness because they can be fine-tuned at a later stage of the process.
- High specificity in the location of DNA rearrangements can be coupled with great flexibility in the exact sequence structures created.
- DNA rearrangements that bring together well-defined functional modules in new combinations are most likely to generate a result useful to the organism.
- Cellular feedback about the functional success of the novel product or process can greatly accelerate its proliferation and subsequent improvement by additional natural genetic engineering steps.

Such currently well-documented and highly integrated evolutionary capabilities were simply unknowable 70, let alone 100 or 150, years ago.

Cellular Regulation of Natural Genetic Engineering

One of the most profound lessons from the past six decades of molecular cell biology is that all aspects of cell functioning and cellular biochemistry are subject to regulation. We have no scientific basis for postulating that genome function and DNA biochemistry are any different in this regard. In other words, we have every reason to expect that natural genetic engineering functions will also be subject to regulation and will not operate in an uncontrolled way, and abundant experimental evidence exists to support this expectation. This field is advancing so rapidly that we now recognize novel regulatory mechanisms that were unanticipated based on what we knew even as recently as the end of the 20th Century.

The best way to approach the question of regulating natural genetic engineering is to return to McClintock's concept of *genome shock* and ask, "What kinds of adverse life experiences trigger genome rearrangements?" The results summarized in Table II.7 indicate that a surprisingly wide variety of stresses activate natural genetic engineering functions. In addition to the predictable effects of DNA damage, we find a considerable number of intercellular signaling, nutritional, physiological, and infectious events that also trigger the activity of genomic change operators.

In many of the examples listed in Table II.7, we have some knowledge of the molecular basis for the natural genetic engineering response. They involve activating cellular response circuits that derepress or positively activate expression of the particular genome restructuring function that responds to the stimulus. For example, starvation-activated transposon-mediated rearrangements in bacteria are known to involve two-component signaling networks, regulatory proteases, and transcription factors such as Crp and RpoS [636–638]. Oxidation- and pheromone-induced retrotransposition in yeast similarly involves well-characterized signaling circuits [639–641].

Table II.7 Various Stimuli Documented to Activate Natural Genetic Engineering (References Appear in the Online Version)

Signal or Condition	Natural Genetic Engineering Function	Organism(s)
Quorum pheromones	DNA release and competence for DNA uptake	Multiple bacteria
Chitin	Competence for DNA uptake	*Vibrio cholerae*
Various stress conditions	Competence for DNA uptake	Gram-positive bacteria
DNA damage	Recombination and mutator polymerases (SOS response)	*Escherichia coli, Bacillus subtilis*, and other bacteria
DNA damage	Prophage excision	*E. coli, B. subtilis*, and other bacteria
DNA damage	Horizontal transfer of integrated conjugative elements (ICE)	Multiple bacteria
DNA damage	ISDra2 transposition	*Deinococcus radiodurans*
DNA damage	Genetic exchange	*Helicobacter pylori*
UV irradiation	Tn10 transposition	*E. coli*
Oxidative stress	SOS responses, prophage induction	Multiple bacteria
Chemical damage	SOS response	*E. coli, Salmonella typhimurium*
Antibiotic	SOS response	*E. coli*
Antibiotic	Competence for DNA uptake	*Staphylococcus aureus*
Antibiotic	Prophage excision	*Staphylococcus aureus*
Antibiotic (beta lactam)	SOS response and horizontal DNA transfer	*Staphylococcus aureus*
Antibiotic	Mutator polymerase	*E. coli*
Tetracycline	CTnDOT excision and conjugal transfer	*Bacteroides sp.*

Table II.7 Various Stimuli Documented to Activate Natural Genetic Engineering (References Appear in the Online Version)

Signal or Condition	Natural Genetic Engineering Function	Organism(s)
Quorum pheromones, plant metabolites (opines)	Conjugal transfer	*Agrobacterium tumefaciens*
Plant phenolics	T-DNA transfer to plant cell	*A. tumefaciens*
Magnetic fields	Tn5 transposition	*E. coli*
Magnetic fields	Tn10 transposition	*E. coli*
Heat shock	F plasmid transfer	*E. coli*
Growth phase	F plasmid transfer	*E. coli*
Genome reduction	Stress-induced IS elements	*E. coli*
Conjugation	ISPst9 transposition	*P. stutzeri*
Sex pheromones	Conjugation agglutinins	*Enterobacter fecaelis*
Nucleic acid precursors	Reduce competence	*Haemophilus influenzae*
Aerobic starvation	Mu prophage activation	*E. coli*
Aerobic starvation	Tn4652 activation	*Pseudomonas putida*
Aerobic starvation	Base substitutions	*E. coli*
Aerobic starvation	Tandem duplications and amplifications	*Salmonella enterica*
Elevated temperature	IS element activation	*Burkholderia sp.*
Elevated temperature and high culture density	IS4Bsu1 element	*B. subtilis*
Adenine starvation	Ty1 retrotransposon activation	*Saccharomyces cerevisiaea*
DNA damage (radiation or carcinogen)	Ty1 retrotransposon activation	*S. cerevisiaea*
Telomere erosion	Ty1 retrotransposon activation	*S. cerevisiaea*

Table 11.7 Various Stimuli Documented to Activate Natural Genetic Engineering (References Appear in the Online Version)

Signal or Condition	Natural Genetic Engineering Function	Organism(s)
MAPK cascade activation during filamentous growth	Ty1 retrotransposon activation	*S. cerevisiaea*
Mating pheromone	Ty3 retrotransposon activation	*S. cerevisiaea*
Mating pheromone	Ty5 retrotransposon activity and transcription	*S. cerevisiaea*
Prion formation	Genome instability	*S. cerevisiaea*
Improper cryopreservation	Ty1 retrotransposition	*S. cerevisiaea*
Oxidative conditions (H_2O_2) mediated by SREBP transcription factor	Tf2 retrotransposon activation	*Schizosaccharomyces pombe*
Nitrogen starvation	LTR retrotransposon transcription	Diatom (*P. tricornutum*)
Aldehyde (decadienal) treatment	LTR retrotransposon transcription	Diatom (*P. tricornutum*)
DNA damage (Mitomycin C)	Transposon and retrotransposon activation	*Drosophila melanogaster*
DNA damage	Alu retrotransposition	*Homo sapiens*
Gamma irradiation	LINE-1 retrotransposition	*Homo sapiens* (human osteosarcoma cells)
Benzpyrene	LINE-1 retrotransposition	*Homo sapiens* (HeLa cells)
Steroid hormones	Mouse mammary tumor virus (MMTV) activation	*Mus musculus*
Plant alarm chemicals	Retrotransposon activation	*Nicotiana tabacum* (tobacco)
Free radical-generating agents, UV-C, or rose Bengal (RB)	Increased homologous recombination, systemically transmitted	*N. tabacum*
Cutting/wounding	Retrotransposon activation	*N. tabacum*
Hydrostatic pressure	MITE DNA transposons	Rice

Table 11.7 Various Stimuli Documented to Activate Natural Genetic Engineering (References Appear in the Online Version)

Signal or Condition	Natural Genetic Engineering Function	Organism(s)
Protoplasting and growth in tissue culture	Transposon and retro-transposon activation	Various plants
Protoplasting and growth in tissue culture	Tos17 retrotransposon activation	Rice
Growth in tissue culture	mPing transposition	Rice
Cell culture growth	1731 LTR retrotrans-poson	*D. melanogaster*
Cell culture growth	LINE-1 element retrotransposition	Mouse cell line
Fungal metabolites	TnT1 retrotransposon	*Nicotiana tabacum*
Chlorine ions (not sodium)	DNA strand breaks and recombination	*Arabidopsis thaliana*
Nickel, cadmium, and other heavy metals	LINE-1 retrotransposi-tion	*Homo sapiens* tissue culture cells
Temperature and day length	Homologous recombina-tion	*A. thaliana*
Helicobacter pylori infection	Adenocarcinoma with microsatellite instability	Human gastric mucosa
Fungal or virus infec-tion	(CT)n microsatellite con-traction	Wheat
Barley stripe mosaic virus (*Peronospora par-asitica*) infection	Increased somatic recom-bination and transposon activation; transmissible systemic response in tobacco	*A. thaliana*, maize, and tobacco
Tobacco mosaic virus and oilseed rape mosaic virus infection	Increased somatic recom-bination (transmissible systemic response)	*N. tabacum*, *A. thaliana*
Temperature	Amplification/reduction in repetitive elements	*Festuca arundinacea* (tall fescue)
Elevation and moisture	BARE-1 retrotransposi-tion	*Hordeum spontaneum* (wild barley)
Heat shock, toxic chemicals	SINE transcription	*Bombyx morii* (silk moth)

Table 11.7 Various Stimuli Documented to Activate Natural Genetic Engineering (References Appear in the Online Version)

Signal or Condition	Natural Genetic Engineering Function	Organism(s)
Various stress conditions	SINE transcription	H. sapiens
Heat shock	B1 SINE transcription	M. musculus
Industrial air pollution	Microsatellite expansion	M. musculus
Particulate air pollution	Germline mutations	Mouse
Chemical mutagens and etoposide	Microsatellite expansion	M. musculus
Diet (extra folic acid, vitamin B12 choline, and betaine)	IAP retrotransposon at Agouti locus (Avy allele)	M. musculus
Lymphocyte differentiation and antigen activation	Activation of VDJ joining, somatic hypermutation, and heavy chain class switching	M. musculus and H. sapiens
Neuronal differentiation and exercise	LINE-1 retrotransposition	M. musculus
Hybrid dysgenesis	P factor transposon	Drosophila melanogaster
Hybrid dysgenesis	I factor non-LTR retrotransposon	D. melanogaster
Hybrid dysgenesis	Hobo transposon	D. melanogaster
Hybrid dysgenesis	Penelope retrotransposon and other transposable elements	Drosophila virilis
Hybrid dysgenesis	Mariner/Tc1, hAT transposons, and gypsy/Ty3 LTR retrotransposons	Medfly (Ceratitis capitata)

The quantitative extent of stimulatory effects on natural genetic engineering can be striking. In the starvation-induced rearrangements that I studied in the 1980s and 1990s, the increase in transposon-mediated events increased by at least five orders of magnitude (that is, by a factor of over 100,000). They went from undetectable in more than 10^{10} bacteria under normal growth conditions to more than once per 10^5 bacteria on starvation plates [642].

In addition to the stimuli summarized in Table II.7, a major class of triggers for natural genetic engineering functions includes atypical sexual encounters in both plants and animals (see Table II.8). These contemporary data confirm what many scientists have asserted repeatedly for over 200 years, from Lamarck in the 18th Century to Stebbins and McClintock in the 20th Century: interspecific hybridization is a major source of evolutionary variability [643] [644].[3]

Table II.8 Genomic Responses to Changes in Ploidy and Interspecific Hybridization in Plants and Animals (References Appear in the Online Version)

Plant Taxon	Genomic Response
Asteraceae (*Compositae*)	Genome expansion and retrotransposon proliferation in sunflower hybrids
	Chromosomal repatterning and the evolution of sterility barriers in hybrid sunflower species
	Rapid chromosome evolution in polyploids
Grasses	Altered methylation patterns and chromosome restructuring in hybrids
Potato	Genome instability in hybrids
Nicotiana spp. (tobacco)	Elimination of repeated DNA in a synthetic allotetraploid
Rice	Extensive genomic variability induced by introgression from wild rice
	LTR retrotransposon movements in rice lines introgressed by wild rice
	Retrotransposon activation following introgression
	Incompatible cross-pollination leading to transgenerational mobilization of multiple transposable elements
	Transpositional activation of mPing in an asymmetric nuclear somatic cell hybrid of rice and *Zizania latifolia* accompanied by massive element loss
Brassica	Rapid genome change in synthetic polyploids
	Large-scale chromosome restructuring

3 This was recently pointed out again in a September 14, 2010, *New York Times* column by Sean Carroll, "Hybrids May Thrive Where Parents Fear to Tread" (http://www.nytimes.com/2010/09/14/science/14creatures.html?ref=science).

Table II.8   Genomic Responses to Changes in Ploidy and Interspecific Hybridization in Plants and Animals (References Appear in the Online Version)

Plant Taxon	Genomic Response
Wheat [645]	Sequence loss and cytosine methylation following hybridization and allopolyploidy
	Rapid genome evolution following allopolyploidy
	Parental repeat elimination in newly synthesized allopolyploids
	Rapid genomic changes in interspecific and intergeneric hybrids and allopolyploids
Arabidopsis	Chromosome rearrangements after allotetraploid formation
	Aneuploidy and genetic variation in the A. thaliana triploid response
	Genomic changes in synthetic polyploids

Animal Taxon	Genomic Response
Drosophila spp.	Increased retrotransposition in interspecific hybrids
Macropus marsupials	Centromere instability in interspecific hybrids
Wallabies	Chromosome remodeling in interspecific hybrids
Mouse	Amplification and double minutes in a hybrid
Rice fish (medaka)	Chromosome elimination in an interspecific hybrid
Odontophrynus americanus (amphibian)	Chromosome instabilities and centromere dysfunction in tetraploids

The most intensively studied examples of natural genetic engineering triggered by atypical sex fall into the category of *hybrid dysgenesis*, first identified in *Drosophila* and later found in mice and wallabies (Table II.7). The term hybrid dysgenesis refers to germline abnormalities (*dysgenesis*) induced by matings between fruit flies of the same species but from different populations (*hybrid*). Hybrid dysgenesis was first noticed in the 1950s and 1960s, when geneticists mated flies from wild populations to laboratory strains established in the early decades of the 20th Century. Frequently, these crosses produced few or no progeny, and those progeny that did result often carried mutations and chromosome rearrangements. I have speculated with a *Drosophila* geneticist friend about how many technicians may have been fired because their crosses did not work as expected in the years before hybrid dysgenesis came to be understood as a predictable phenomenon.

Initially confusing, the chromosomal basis of several distinct hybrid dysgenesis systems was identified in the 1980s. The causes are active transposons or retrotransposons in the wild flies that are missing or inactivated in the older populations of laboratory flies [646–648]. These mobile genetic elements are active specifically in the developing germline tissues of the hybrid flies, where they cause chromosome breakage, high rates of transposition, and chromosome rearrangements. As with the bacterial example mentioned earlier, the range of activation can be impressive. From being undetectable within each of the parental populations, the probability of genetic change in the hybrids can exceed 100% (in other words, more than one change can be found in every progeny of a dysgenic mating).

Hybrid dysgenesis is important to understand because it provides a way to make sense of the population genetics needed for fixation of sudden large-scale genome rearrangements. The fact that transpositions and other genome changes occur before meiosis during the mitotic development of the germline means that multiple sperm or egg cells carrying these changes arise at the meiotic conclusion of germline development. The resulting gametes can transmit a constellation of hybrid dysgenesis-related changes [463, 649]. Because these gametes can participate in several fertilization events, they can produce a small population of individuals, each carrying the same alterations to genome structure. The individuals in this population then can interbreed to fix a novel genome architecture.

Genetic analysis indicated that, in addition to the mobile genetic elements, hybrid dysgenesis has another component called *cytotype*. This referred to the capacity of fertilized egg cells to control the activity of the mobile elements [646, 650]. Only in the last few years has the molecular nature of cytotype control been deciphered. The control depends on the tendency of eukaryotic cells to silence transposons and retrotransposons epigenetically by incorporating them into silent chromatin. Such silencing is directed by small RNA molecules, which are called *piRNAs* because they bind a *Piwi* protein component of the germline-specific epigenetic silencing mechanism [651–654]. The piRNAs are processed from transcripts of specialized genetic loci that incorporate many different fragments of mobile elements and viruses that have entered the *Drosophila* genome in the past [655, 656]. These fragments provide the sequence specificity needed to direct the epigenetic silencing apparatus to the hundreds of dispersed repeats of mobile elements in the genome. If a *Drosophila* egg cell genome lacks the piRNA sequences corresponding to a mobile element, it cannot silence expression of that element after fertilization, and hybrid dysgenic events occur in germline development.

Note that the piRNA epigenetic control system must depend on a DNA acquisition apparatus in *Drosophila* (and in other organisms that have equivalent RNA-directed silencing regimes). This apparatus can capture fragments of mobile elements, viruses, and other genome invaders and then insert them into appropriate piRNA-encoding loci. We do not yet understand how this newly discovered genome surveillance *cum* natural genetic engineering system operates. However, such "genomic immunity" is not limited to animals or even to eukaryotes (Table II.9). A similar system has recently been observed in prokaryotes, whose genomes have special structures called CRISPRs.

Table II.9 RNA-Based Defense Against Viruses and Plasmids Identified in Archaea, Bacteria, Fungi, and Plants (References for Individual Species Appear in the Online Version)

Archaea

Crenarchael thermophiles, *Pyrococcus furiosus*, *Sulfolobus islandicus*, *Sulfolobus solfataricus*

Bacteria

Lactic acid bacteria, *C. diptheriae*, *E. coli*, *Leptospirillum* group II bacteria, *Pseudomonas aeruginosa*, *Streptococcus thermophilus*, *Streptococcus mutans*, *Staphylococcus epidermidis*, *Thermus thermophilus*, *Vibrio cholerae*, viridans streptococci, *Xanthomonas oryzae*, *Yersinia pestis*

Yeast and Fungi

Saccharomyces cerevisiae (budding yeast), *Schizosaccharomyces pombe* (fission yeast), *Neurospora crassa* (bread mold), *Aspergillus nidulans* (fungus)

Plants

Zea mays (maize), Rice (both monocots); *Arabidopsis thaliana*, Tomato, *N. tabacum*, Grapevine, *Craterostigma plantagineum* (blue gem, resurrection plant) (all dicots); Conifers (gymnosperms)

CRISPR stands for *clustered regularly interspaced short palindromic repeats*. CRISPRs were initially identified by bioinformatic analysis that annotated their repetitive structures [657]. It turns out that the active components are actually the spacer elements between the CRISPR repeats. As in piRNA-encoding loci, the CRISPR spacers accumulate fragments of viruses, plasmids, and other genome invaders [658, 659]. Although prokaryotes do not seem to possess the same kinds of epigenetic silencing machinery as eukaryotes, protein sequence motifs and recent experiments indicate that spacer transcripts produce small silencing RNA molecules (siRNAs) that prevent virus replication [660–664]. In addition, experiments demonstrate that bacteria sensitive to particular viruses can become resistant by incorporating viral sequence fragments as CRISPR spacers [659, 665]. Currently, we are as ignorant about how CRISPRs incorporate fragments of invading molecules as we are about piRNA loci. However, the fact that the incorporation process can be documented experimentally in easily studied bacteria means that mechanistic understanding will not be long in coming.

The CRISPR system illustrates an important point about the history of science. CRISPRs demonstrate the real-time operation of a process long dismissed by traditional geneticists and evolutionists as disproven, once and for all, by the famous Luria-Delbrück fluctuation test of 1943 [666]. Dozens of textbooks explained that the culture-to-culture variations observed by Luria and Delbrück proved that infection could not induce resistance. In fact, what Luria and Delbrück demonstrated was that mutations conferring resistance to a virus that is invariably lethal immediately upon infection do occur prior to selection (as they must). They never could disprove the operation of a CRISPR or other infection-triggered resistance mechanism for other viruses, such as temperate bacteriophages. The incorporation of fragments from invading DNA elements for the purpose of self-defense (the CRISPR system has been described as a genomic immune system [667]) is a precise example of the kind of dedicated, nonrandom, beneficial change specifically excluded by generations of evolutionary theorists [3].

The overgeneralization of conclusions from the Luria-Delbrück experiment further illustrates an unfortunate tendency in evolutionary studies (and in science in general) to make sweeping assertions about what can or cannot happen based on relatively limited experimental observations. The common impulse is to declare "impossible" what does not agree with the assumptions or prejudices of a particular school of thought [31]. Because it is the business of science to turn what was once thought to be impossible into reality, noting the perils of excessive generalization reminds us to keep our ideas fresh, creative, and inclusive rather than rigid and exclusive.

In keeping with this appeal for fresh ideas, we have to recognize that epigenetic regulation of mobile genetic elements (and perhaps of other natural genetic engineering functions) opens new possibilities for understanding the connections between life history events and genome change. Any processes that perturb the epigenetic markings of the genome are candidates for genome-destabilizing agents. Indeed, events such as interspecific hybridization, changes in chromosome number, and microbial infection all lead to alterations of epigenetic formatting (see Table II.10) as well as genomic instabilities in plants and animals (see Table II.8). The idea is currently circulating in the scientific literature that the *epigenome* (the collective suite of chromatin formats that regulate genome functioning) serves as a major interface between individual life experience and genome expression [262, 263, 668]. Because epigenetic formats are heritable, this idea suggests that the genome can be destabilized for one or more generations when populations undergo major ecological disruptions, and episodes of transgenerational destabilization have been reported (see Table II.8).

Table II.10 Life History Events That Alter the Epigenome (DNA Methylation and Chromatin Formatting) (References Appear in the Online Version)

Event (Organism)	Results
Polyploidy (plants)	Histone acetylation changes
Allopolyploidy (plants)	Changes in methylation patterns of mobile elements
Synthetic allotetraploids (*A. thaliana*)	Remodeling of DNA methylation, phenotypic, and transcriptional changes
Interspecific hybrids (plants)	Altered DNA methylation patterns; phenotypic and epigenetic variability
Fungal infection (tobacco)	Transgenerational instability
Introgression from *Zizania latifolia* into rice	Extensive alterations in DNA methylation

Table II.10 Life History Events That Alter the Epigenome (DNA Methylation and Chromatin Formatting) (References Appear in the Online Version)

Event (Organism)	Results
Incompatible cross-pollination (rice)	Transgenerational epigenetic instability
Tissue culture growth (rice)	Altered mPing transposon cytosine methylation
N. tabacum tissue culture	Gradual and frequent epigenetic reprogramming of invertedly repeated transgene epialleles
Immortalized *A. thaliana* cell suspension culture	Euchromatin DNA hypermethylation and DNA hypomethylation of specific transposable elements
Antibiotics and tissue culture (*N. tabacum*)	Genome-wide hypermethylation
Rice plants subjected to space flight	Heritable hypermethylation of TEs and other sequences
Interspecific hybrids (mouse)	Placental DNA methylation changes
Interspecific hybrid (mouse)	Methylation perturbations in retroelements
Interspecific hybrids (*Peromyscus* mice)	Genomic imprinting disrupted
Interspecific hybrids (wallabies)	Loss of retroelement methylation
X irradiation (humans)	Transgenerational cancers and modifications in DNA methylation
Particulate air pollution (mouse)	DNA global hypermethylation dietary methionine
Wolbachia endosymbiosis in *Drosophila* males	Sperm chromatin remodeling; cytoplasmic incompatibility
Wolbachia endosymbiosis in leafhopper (*Zyginidia pullula*) males	Feminization, sterility, female-specific DNA methylation patterns
Bacterial infection (human)	Histone modifications and chromatin remodeling (particularly immune cells)
Bacterial infection (mice)	DNA hypermethylation
Helicobacter infection (human)	LINE-1 hypomethylation
Helicobacter infection (human)	Aberrant or hypermethylation of CpG islands
Campylobacter rectus infection of the placenta (human)	DNA methylation and histone modification changes

Targeting of Natural Genetic Engineering within the Genome

There is one last area where the traditional assumptions about genetic change have been shown to be unrealistically restrictive. That is the question of targeting changes to specific regions in the genome. Conventional wisdom and the vast majority of evolutionists assert that there is no way natural genetic engineering functions can "choose" where to operate within the genome. This was a topic of active debate in 1988 when some adaptive mutation experiments were initially overinterpreted in neo-Lamarckian terms [2, 669, 670].

Despite interpretive errors in the Lamarck *vs*. Darwin debate, *a priori* denials of the capacity for functional targeting of biochemical changes to DNA should be jarring to molecular biologists. We have over 50 years of investigation into the molecular basis of how cells regulate transcription, and all biologists agree that the transcription apparatus can be directed to specific, functionally appropriate sites in the genome. The reason for the denial in the case of mutation probably has to do with a continuing influence of the late 19th Century philosophical notion that "germ plasm" inheritance has to be isolated from the soma [671]. But in the 21st Century, when we know about transcriptional regulation, signal transduction from the cell surface to the genome, and the operation of natural genetic engineering in the germline, it is time to abandon this mistaken doctrine.

It is difficult (if not impossible) to find a genome change operator that is truly random in its action within the DNA of the cell where it works. All careful studies of mutagenesis find statistically significant nonrandom patterns of change, and genome sequence studies confirm distinct biases in location of different mobile genetic elements. These biases can sometimes be extreme, as in the targeting of *S. cerevisiae* LTR retrotransposon insertions into regions just a few base pairs upstream of RNA polymerase III transcription start sites [672–674]. In many cases, we have some understanding of the molecular mechanisms and/or functional significance of the observed preferences (see Table II.11).

There is nothing mechanistically surprising about targeting for natural genetic engineering. The mechanisms are the same as those used to target transcription and homologous recombination, and they also operate by coupling genome changes to other cellular functions, such as transcription or DNA replication. The list of mechanisms targeting natural genetic engineering in Table II.11 reads like the basic elements found in a molecular biology textbook: (i)DNA-DNA homology, (ii) RNA-DNA homology, (iii) sequence-specific binding of proteins to DNA, (iv) structure-specific binding of proteins to DNA, (v) protein-protein interactions, and (vi) protein recognition of specific chromatin configurations.

One intriguing case where targeting has been known for over 25 years, but the underlying mechanism has not yet been fully clarified, is the phenomenon of P element *homing* [675–680]. P elements are DNA transposons that induce hybrid dysgenesis in the fruit fly. They also insert into the *Drosophila* germline genome at high frequency after microinjection into embryos [681, 682]. For this reason, cloning DNA fragments into P elements has been developed as the standard method for creating *transgenic* fruit flies (flies with external DNA inserted into the genome) [270, 683, 684]. In the course of using this method, a number of investigators have observed that the insertion specificity of the engineered P elements changes according to the added DNA. The changes are functionally significant; the inserted DNA directs the engineered P element at high frequency (~20 to 50%) into regions of the genome where proteins that recognize the inserted DNA play an active role in regulation. For example, insertion of a binding sequence for the transcriptional regulator Engrailed targets a large fraction of insertions to chromosomal regions where Engrailed is known to function [676].

P element targeting has nothing to do with sequence homology, because the homing insertions go to many different sites within a large region around the corresponding genome sequence (such as the transcription factor binding site). The most plausible hypothesis to explain P element homing derives from the obvious point that transposon insertions require that the donor element and target site be brought physically close together in the nucleus [678]. According to this hypothesis, a transcription factor binding site or chromatin formatting site inserted into a P element causes that element to localize

frequently to a nuclear subcompartment where the cognate transcription factor or chromatin formatting function is active. That colocalization is what biases insertion events to the corresponding regions
of the genome.

The subnuclear localization hypothesis for P element homing has
the dual virtues of conferring functional specificity while permitting
flexibility in target site selection. This combination of larger-scale
specificity and finer-scale flexibility is one of the evolutionary principles exemplified by the mammalian adaptive immune system. We will
return to this point in Part IV when we consider some of the conceptual implications of evolution by natural genetic engineering. For
now, let us simply accept as established the point that cells have the
ability to target genome change in a wide variety of ways that make
sense from a molecular biology perspective (see Table II.11).

Table II.11 Examples of Targeted Natural Genetic Engineering (References
Appear in the Online Version)

Example	Observed Specificity (Mechanism)
DNA import and export	Special DNA uptake signals. *oriT* sites for initiating conjugal transfer replication.
Homologous recombination	Special sequences stimulating DS breaks and other biochemical events in homologous exchange.
Transposon insertions at special DNA structures	Insertion at REP palindromes (transposase specificity), DNA replication forks (interaction with processivity factor).
IS200/IS605 family target site selection	DNA sequence homology.
IS911 target site selection	InsAB transposase binding to specific DNA sequences. Regulated by synthesis of InsA transposase without specificity.
Cassette replacement/conversion in antigenic variation	DNA sequence homology at cassette boundaries.
Site-specific recombination (phase variation, antigenic variation, insertions and excisions)	Protein recognition of DNA sequence. Protein-protein interaction.
Diversity-generating retroelements	Localized mutagenesis at duplicated segment of coding region. Reverse transcription, RNA-DNA sequence homology.

Table II.11 Examples of Targeted Natural Genetic Engineering (References Appear in the Online Version)

Example	Observed Specificity (Mechanism)
Mating type cassette switching (*S. cerevisiae, S. pombe, Kluyveromyces lactis*)	Protein recognition of DNA sequence (endonuclease or transposase cleavage at unique site), DNA sequence homology at cassette boundaries.
Hermes transposon in *S. cerevisiae*	Preferential insertion in nucleosome-free regions.
Immune system V(D)J joining	Cleavage at specific recombination signal sequences (recognition of RSSs by RAG1+2 transposase). Flexible joining by nonhomologous end-joining (NHEJ) functions.
Immune system somatic hypermutation	5' exons of immunoglobulin sequences (transcriptional specificity determinants), DIVAC element to suppress repair.
Immune system class switching	Lymphokine-controlled choice of switch region transcription (promoter activation).
Budding yeast (*S. cerevisaea*) retroviral-like elements Ty1 through Ty4	Strong preference for insertion upstream of RNA polymerase III initiation sites (protein-protein interaction of integrase with RNA polymerase III factors TFIIIB and TFIIIC).
Budding yeast retroviral-like element Ty1	Preference for insertion upstream of RNA polymerase II initiation sites rather than exons.
Budding yeast retroviral-like element Ty5	Strong preference for insertion in transcriptionally silenced regions of the yeast genome (protein-protein interaction of integrase targeting domain (TD) with Sir4 silencing protein). Regulated in response to stress by modulation of integrase TD protein phosphorylation.
Fission yeast (*S. pombe*) retroviral-like elements Tf1 and Tf2	Insertion almost exclusively in intergenic regions (>98% for Tf1). Biased toward PolII promoter-proximal sites, 100 to 400 bp upstream of the translation start, by protein-protein interaction with transcription activators. Prefers chromosome 3.
MAGGY (fungal Ty3/gypsy family) retrotransposon	Targeting to heterochromatin by chromodomain in integrase protein.

Table II.11 Examples of Targeted Natural Genetic Engineering (References Appear in the Online Version)

Example	Observed Specificity (Mechanism)
Dictyostelium discoideum non-LTR retrotransposon TRE5-A	Insertion upstream of tRNA sequences by protein-protein interactions with RNA Pol III transcription factors.
Rapidly expanding mPing transposons in rice	Insertion upstream of coding sequences.
Drosophila ZAM LTR retrotransposons	Site-specific insertions by protein-DNA recognition.
Murine Leukemia Virus (MLV)	Preference for insertion upstream of transcription start sites in human genome. Role for IN (integrase) and GAG proteins.
HIV, SIV	Preference for insertion into actively transcribed regions of human genome. Role for IN (integrase) and GAG proteins. HIV integrase interaction with LEDGF/p75 transcription factor.
Gammaretroviral (but not lentiviral) vectors	Insertion at transcription factor binding sites. 21% recurrence rate at hotspots.
Drosophila gypsy retrovirus	Site-specific insertion into *Ovo* locus regulatory region guided by *Ovo* protein binding sites.
Drosophila P-factors	Preference for insertion into the 5' end of transcripts.
Engineered *Drosophila* P-elements	Targeting ("homing") to regions of transcription factor function by incorporation of cognate binding site. Region-specific.
HeT-A and TART retrotransposons	Insertion at *Drosophila* telomeres.
R1 and R2 LINE element retrotransposons	Insertion in arthropod ribosomal 28S coding sequences (sequence-specific homing endonuclease).
Group I homing introns (DNA-based)	Site-specific insertion into coding sequences in bacteria and eukaryotes (sequence-specific endonuclease).
Group II homing introns (RNA-based)	Site-specific insertion into coding sequences in bacteria and eukaryotes (RNA recognition of DNA sequence motifs, reverse transcription).
Group II intron retroelements	Insertion after intrinsic transcriptional terminators.

Reviewing What Cells Can Do to Rewrite Their Genomes over Time

Cells operate at roughly three different *biological* time scales:

1. Within a single cell cycle
2. Over a finite number of cell cycles (as in plant or animal morphogenesis), and
3. Over the countless cell cycles of evolutionary time.

At each of these three time scales, cells reorganize their genomes, thereby inscribing information that influences all aspects of genome function.

Within the cell cycle, most of this inscription occurs by forming transient nucleoprotein complexes that carry out replication, transcription, repair, and physical movement of the genome at each stage of cell growth and division. Over several cell cycles, at least one main form of inscription is epigenetic modification by imprinting and chromatin formatting. These epigenetic inscriptions are heritable through many normal cell divisions, but they are also subject to modification by chromatin reformatting and imprinting systems, particularly during meiosis and gamete formation. In a number of cases just described, such as phase and antigenic variation (see Table II.5) and antibody formation (see Appendix II.2 online), cells also rewrite intermediate term information by DNA structural changes.

Over evolutionary time, cells write new information into their genomes by the numerous natural genetic engineering processes described earlier and by two further modes of genome change that we will discuss in Part III: horizontal DNA transfer and symbiotic cell fusions. These processes lead to long-term changes in the DNA structure at all levels of genome complexity. Natural genetic engineering can act massively throughout the genome within a single cell generation, as illustrated by ciliate protozoa. Natural genetic engineering frequently functions in a targeted fashion at particular locations within the genome, and such targeted changes have been incorporated into adaptively useful examples of vegetative and somatic evolution. Because of the relationship between the cellular regimes that control the activity of natural genetic engineering functions and sensory networks, we see clearly how episodes of genome restructuring can be linked to moments of adaptive crisis, such as starvation or population collapse.

Part III discusses how the action of natural genetic engineering functions has generated many important features found in present-day whole genome sequences. We just need to think of the more than three million transposition and retrotransposition events in our own genome's history to recognize a striking example [382]. You will also see how molecular analysis of the evolutionary process has opened up numerous conceptual possibilities by revealing that life is more diverse than we imagined, by demonstrating that vertical transmission from ancestors is not the only way that heritable characters are acquired, and by documenting the power of cell fusions as a force for generating novel life-forms. Therefore, let us turn to the DNA record to learn some of its incontrovertible lessons.

Evolutionary Lessons from Molecular Genetics and Genome Sequencing

In this part:

- Antibiotic resistance and horizontal DNA transfer
- The modular and duplicative nature of protein evolution
- Molecular taxonomy and the discovery of a new cell type
- Symbiogenesis and the origin of eukaryotic cells
- Natural genetic engineering and evolutionary genomic innovation
- Use and reuse of evolutionary inventions
- What makes a man different from a mouse?
- Whole genome doubling at critical stages of evolutionary innovation and divergence
- Reviewing what the DNA record reveals about cell activities over evolutionary time

The advent of molecular genetics and genome sequencing was a major step forward in evolutionary science. Examining the DNA record made it possible to subject traditional evolution theories to rigorous empirical testing. Do the sequences of contemporary genomes fit the predictions of change by "numerous, successive, slight variations," as Darwin stated [1], or do they contain evidence of other, more abrupt processes, as numerous other thinkers had asserted [644, 685–687] **[643, 688]**? The data are overwhelmingly in favor of the saltationist school that postulated major genomic changes at key moments in evolution. Only by restricting their analyses to certain classes of genomic DNA, such as homologous protein coding sequences, can conventional evolutionists apply their gradualist models. Moreover, we will see from genome sequencing that protein evolution itself often proceeds in relatively large steps. Contrary to the views of Linnaeus and Darwin, nature does indeed make leaps, and we now have molecular evidence of how some leaps occurred.

Antibiotic Resistance and Horizontal DNA Transfer

Ever since World War II, we have been conducting a massive, real-world evolutionary experiment by the widespread application of antibiotics in medicine and animal husbandry. The introduction of massive amounts of chemically distinct antibacterial agents into the human and animal environment constituted a major alteration of the selective landscape. Predictably, the bacteria have responded by evolving resistance, and today the continued effectiveness of our antibiotic arsenal is in doubt [689]. Did we underestimate these small but ubiquitous cells?

In the 1950s, at the time Watson and Crick articulated the structure of DNA, there was a coherent *and experimentally verified* theory of how bacteria acquired antibiotic resistance by mutation. Individual mutations would modify cell structures so that they became less sensitive to or made the bacteria less permeable to a particular antibiotic. In the laboratory, bacterial geneticists succeeded in isolating the predicted mutations. For a few antibiotics, such as streptomycin, a single mutation could modify the ribosomal target and produce resistance to a high level [690]. However, for most other antibiotics, single mutations led to small increments of resistance so that multiple successive mutations were needed to achieve clinically significant resistance, in accord with traditional views [397] [691]. Generally, increased resistance by mutation came at a cost in a slower bacterial growth rate. So it was assumed that the evolution of bacterial antibiotic resistance by mutation would be retarded by the growth defects of the mutant bacteria. Acting at least in part on this assumption, in 1967 the U.S. Surgeon General asserted that "the war against infectious diseases has been won" [692].

There was only one problem with this experimentally confirmed theory: It was wrong. It did not recognize the genomic capabilities of the target bacteria. In clinical settings, bacterial antibiotic resistance spread rapidly, and the introduction of new types of antibiotics was quickly followed by the emergence of strains resistant to multiple antibiotics at once. By the early 1960s, studies carried out largely in Japan had clarified this rapid evolutionary process. Rather than becoming resistant by mutation, bacteria became resistant by acquiring transmissible genomic elements (*multidrug resistance plasmids*) encoding a variety of novel biochemical activities that blocked antibiotic action [693, 694] **[695]**. Some of these new functions inactivated antibiotics by hydrolysis (penicillins) or chemical modification (streptomycin, chloramphenicol). Other functions pumped the antibiotic out of the cell (tetracycline). And still other functions chemically modified cellular targets (erythromycin) or substituted drug-resistant activities for drug-sensitive cellular ones (sulfanilamides, trimethoprim) [696–698].

Both the genetic and biochemical virtuosities displayed by the multiply resistant bacteria were a major surprise for microbiologists, biochemists, and geneticists. For many decades, they went ignored by evolutionists, who were far more interested in large eukaryotes than in microbes. The traditional focus on larger organisms notwithstanding, the study of transmissible antibiotic resistance turned out to hold three major evolutionary lessons that have proven to be applicable in all realms of life:

- **Horizontal transfer of DNA molecules** is widespread in nature. Contrary to traditional theories, it is now well documented that all prokaryotes and many eukaryotes acquire novel genomic segments and biochemical functions from other, often unrelated cells rather than exclusively by vertical inheritance from progenitors. Such theories of exclusively vertical transmission were the basis of Darwin's "tree of life" diagram at the end of *Origin of Species* [1].

  Table III.1 lists various documented occurrences and mechanisms of horizontal DNA transfer between similar or distinct types of organisms. Among these mechanisms are the DNA import and export systems mentioned in Part II. Another important group of DNA transfer vectors are infectious agents: viruses, prokaryotic cells, and eukaryotic parasites (large and small). Many of these agents can infect widely different cell types and can combine DNA from taxonomically distant realms of life. This combinatorial function is dramatically illustrated by the newly discovered family of *nucleocytoplasmic large DNA viruses*, whose megabase-sized genomes contain segments originating from both realms of prokaryotes as well as from various eukaryotes [699–702].

• **Integration and engineering of horizontally transferred DNA** is a key feature of evolutionary change. In the course of studying multidrug resistance determinants, bacterial geneticists uncovered several natural genetic engineering processes involved in joining different resistance determinants. The two most important are DNA transposition and site-specific recombination, frequently combined within a single composite mobile genetic element.

Transposition events can insert one or more antibiotic resistance determinants into a transmissible DNA molecule (plasmid or viral) that already encodes other resistances. In addition, many multiple resistance determinants are constructed from individual coding sequence cassettes by site-specific recombination into structures called *integrons* (see Table II.6). Integrons encode an *integrase* protein that inserts and excises cassettes in much the same way as temperate phage genomes are inserted into and excised from the bacterial chromosome [703–705].

Intriguingly, at least one integron recombinase has recently been found to work on single-stranded rather than double-stranded cassettes. The single-strand recombinase activity links the integration event to conjugal transfer of DNA or DNA uptake in transformation; in both cases, the newly introduced DNA is single-stranded [706].

Integrons have been found in plasmids and inside transposons. Apparently, the formation of a mobile multiple antibiotic resistance determinant follows a process of sequential assembly from individually mobile subcomponents. Integrons have also been found in extended form (called *superintegrons*) in the chromosomes of several prokaryotic species. These superintegrons contain dozens of cassettes permitting complex adaptations [707–709].

- **No inviolable taxonomic barriers** exist for horizontally transferred DNA (see Table III.1). Prokaryotes and eukaryotic microbes in a common environment freely share DNA. Some bacteria regularly transfer DNA to plants during tumorigenesis or symbiosis. Endosymbiotic bacterial sequences have been shown to integrate into the host insect genome. Pathogenic bacteria can pick up host DNA. These observations demonstrate the absence of fundamental taxonomic barriers to horizontal DNA transfer. Thus, contemporary evolutionary theories have to incorporate horizontal transfer of multiple coding sequences from any realm of life as a basic mode of genome change. Note that our knowledge of the true extent of DNA acquired in this manner during evolution is limited to relatively recent events because sequence divergence over time obscures the evidence identifying horizontal transfer.

Table III.1   Examples of Intercellular and Interkingdom DNA Transfer (References Appear in the Online Version)

Horizontal Transfer Mode	Documented Transfers
Uptake of environmental and liposomal DNA	*Bacteria–bacteria*
	Archaea–archaea
	Algal transfection
	Plant–*bacteria*
	Plastid transfection
	Mammalian cell transfection, lipofection
	Plant protoplasts
Conjugal transfer	*Bacteria–bacteria*
	Archaea–archaea
	Bacteria–archaea
	*Bacteria–*yeast
	*Bacteria–*plant
Viral transduction and GTAs (gene transfer agents)	*Bacteria–bacteria*
	Archaea–archaea
	*Bacteria–*plant
	Animal cell–animal cell
	Animal cell–virus

Table III.1   Examples of Intercellular and Interkingdom DNA Transfer (References Appear in the Online Version)

Horizontal Transfer Mode	Documented Transfers
Direct fusion of cells or protoplasts that lack a rigid outer covering	*Bacteria–bacteria*
	Mammal–mammal
	Plant–plant
Parasitic or endosymbiotic association	Plant–fern
	Plant–plant
	Bacteria–invertebrate
Undetermined mechanism	*Archaea–bacteria*
	Bacteria–fungus
	Bacteria–protist
	Archaea–protist
	Protist–protist

Prokaryotes frequently use horizontal transfer to tailor their genomes for particular ecological opportunities. They contain many specialized genetic elements, such as conjugative plasmids and conjugative transposons, that transfer extended genome segments [710, 711]. In the early 1980s, two little-known scientists from Montreal, Sorin Sonea and Maurice Panisset, published *A New Bacteriology* [712]. The premise of this unconventional book is that bacteria do not have fixed specific genomes but instead share a vast genome distributed across multiple cells and virus particles. When a specific ecological opportunity appears, the bacteria form a specialized cell to exploit the opportunity by incorporating the appropriate DNA fragments from this distributed genome.

Genome sequence analysis of individual bacterial genomes supports the Sonea-Panisset idea [713, 714]. The sequences show that specialized ecological functions, such as pathogenesis, symbiosis, and metabolism of sparsely distributed chemicals, are frequently encoded in structures called *genomic islands* [439, 715]. These islands can be many thousands of kilobase pairs in length and generally contain DNA demonstrably different in origin from the rest of the genome (that is, of a distinct nucleotide composition). Moreover, genomic islands typically are flanked by sequence markers characteristic of site-specific recombination or transposition events, indicating their recent integration into the genome. Some genomic islands have recently been found packaged into virus-like particles that can infect bacteria from a different genus than the cell of origin [716–718].

Additional support for the idea of a distributed prokaryotic genome comes from the recent practice of *metagenomics*, which involves taking environmental samples and sequencing the DNA without culturing the organisms it contains [719] [720–723]. Metagenomics provides a snapshot of the total DNA content of a particular ecological niche, ranging from seawater and acid mine drainage to the contents of an animal or human intestine [724–727]. The results of this environmental analysis reveal that we currently know about far less than 1% of the extant microbes on Earth from isolation in laboratory culture. We also know that each new environment yields previously unknown DNA sequences encoding proteins new to biochemistry [728–730]. Many of these newly discovered sequences come from viral particles and thus demonstrate that the *virosphere* provides a major reservoir of genetic novelty [731–733].

The Modular and Duplicative Nature of Protein Evolution

The 2001 *Nature* report of the draft human genome contained two important figures illustrating what genome sequencing had taught us about protein evolution [382]. Using transcription factors and chromatin binding proteins as examples, the figures showed that these classes of proteins did not evolve one amino acid at a time. Instead, the two classes of protein "shuffled" and "accreted" copies of functional protein segments called *domains* as eukaryotes progressed from yeast through nematode worms and *Drosophila* fruit flies to mice and human beings. In other words, proteins diversify through a process of acquiring, amplifying, and rearranging coding sequences for subprotein structures that may be dozens or hundreds of amino acids in length.

Early thinking about protein evolution focused on the single amino acid as the basic unit of variation, corresponding to a single nucleotide change in the coding DNA [734, 735]. One of the most important advances sequence analysis brought to the study of protein evolution was the recognition that proteins share defined segments (*domains*) more widely than they share overall structure [736] [737–740]. This recognition coincided with genetic engineering experiments in the laboratory demonstrating that a domain from one protein could be joined to a domain from another and that both domains would retain their functionality. One of the earliest examples involved switching DNA sequence recognition domains in transcription factors that changed the specificity of regulation [741] [742]. Both the sequencing results and genetic engineering experiments led to the view of proteins as composite structures built as systems of structurally and functionally autonomous domains. Thus, a transcription factor such as the LacI repressor is considered to be composed of DNA binding, inducer binding, and protein-protein interaction domains [743] (see the illustration in Appendix I.1 online). Indeed, when a new protein sequence has been determined from genomic DNA today, it often proves more instructive to search the domain database for matches to infer functionalities than it does to search the intact protein database. Domain comparison searches are now routine for protein analysis [744, 745] (http://pfam.sanger.ac.uk).

The concept of most proteins as systems of domains exemplifies the new combinatorial thinking frequently emphasized in this book. It makes good sense *a priori* to expect that a protein will make a successful functional change by acquiring an existing intact binding or catalytic capability. Intuitively, this has a far higher probability of proving effective than does a random process of changing one amino acid at a time and gradually selecting modest improvements in catalysis or binding specificity. In many cases, existing sequences do not provide suitable starting material for evolving new functions one amino acid at a time, because those novel functions require entirely different polypeptide structures. But this restriction does not apply to the process of acquiring an entire new domain, which already comes appropriately configured. The fact that artificial protein evolution in the laboratory often works far better by domain-swapping methods than by localized mutagenesis is yet another indication that the former is a more effective protein innovation strategy than individual or multiple independent amino acid changes [746].

The systems view of proteins implies that they evolve by natural genetic engineering rather than by localized mutation. Is there experimental evidence that the requisite processes for swapping domain-coding sequences actually occur in living cells? Some of the earliest molecular genetics experiments in bacteria involved the formation of hybrid proteins by deletion events that eliminated termination signals and joined two coding sequences into one (for example, [747]). My late colleague Malcolm Casadaban developed a generalized *in vivo* technique using a DNA transposon that could fuse any *E. coli* protein coding sequence to the enzymatically active domains of LacZ beta-galactosidase [748]. Mammalian tissue culture experiments have demonstrated the domain-swapping capabilities of non-LTR retro-transposons through retrotransduction, either of upstream sequences (SVA elements) or downstream sequences (LINE elements) [522, 749, 750]. In addition to domain swapping by retrotransduction, genome sequences in plants and animals have begun to document protein-coding regions where new exons have been incorporated by different classes of DNA transposons (so-called "Pack-MULEs" in rice and helitrons in maize [213, 751–754]). So the capacity of living cells to carry out the requisite natural genetic engineering operations for protein evolution by domain swapping is unequivocally established.

It has become common in the literature to refer to groups of proteins with similar functions by one or more key domains they share. You will see an example later when we talk about the *homeodomain* proteins that have proven so central to contemporary ideas about animal development. For proteins to share domains, of course, the DNA sequences encoding the domains have to be amplified and then shuffled by natural genetic engineering processes. This sequence of events is key to protein evolution, but it has not been systematically investigated in the laboratory (beyond showing that processes such as retrotransduction are feasible). It is to be hoped that some enterprising young geneticist will devise an experimental protocol for analyzing the full potential of domain amplification in real time.

In addition to individual domains, whole proteins are commonly amplified into *protein families* [755–758]. This process is as common in prokaryotes as in eukaryotes [759, 760]. Typically, the amplified proteins carry out a key adaptive function, such as fatty acid metabolism in

Mycobacterium species (the tuberculosis bacterium and its relatives) [757, 761, 762]. The nature of the proteins that have been iterated to form families is frequently a good taxonomic as well as ecological indicator; each group has its own characteristic protein family expansions. In some cases, such as the families of olfactory receptor proteins in mammals (the largest protein family we have [763] [764–766]), it is possible to infer at least one amplification mechanism. In this particular case, reverse transcription of messenger RNAs appears to have played a role, because some of the receptor coding sequences in the DNA genome lack introns found in coding sequences for related receptors [509].

Molecular Taxonomy and the Discovery of a New Cell Type

One of the most important discoveries in 20th Century molecular evolution studies was the demonstration of a third, unexpected type of living cell. In the early 1970s, Carl Woese and his collaborators decided to use the nascent technology of nucleic acid sequencing to put taxonomy on a firm molecular basis [767]. To do this, they had to choose a particular cell molecule as their taxonomic indicator. Woese decided to use the RNA of the small ribosomal subunit for several reasons:

- The ribosome is a basic organelle common to all cells as the site where protein polymerization occurs by translating RNA code into amino acid sequence.

- The ribosomes of all cells have similar but distinguishable structures, indicating that changes occurred in evolution but were restricted by the need to maintain functionality in such a basic cell constituent.

- Ribosomes are the most abundant of cell organelles and thus provide a ready supply of RNA for nucleotide sequence determination (more advanced for RNA than for DNA when Woese began his studies).

- The size of the small ribosomal subunit RNAs (~1,500 nucleotides in bacteria, ~1,900 nucleotides in eukaryotes) was at the practical limit of contemporary methods of sequence determination [768].

When Woese carried out his ribosomal RNA comparisons on a wide variety of prokaryotic and eukaryotic cell types, he was in for a big surprise. Rather than two different families of related RNA sequences, there were three **[769]** [770, 771]. Not all prokaryotic cells had related ribosomal RNA molecules. Prokaryotic cells, previously thought to be a single class, resolved into two separate groups as distinguishable from each other in their ribosomal RNA sequences, as both were from eukaryotic cells. A fundamentally distinct type of cell had been discovered. Because most early examples of this new cell type were isolated from environments considered similar to the early Earth (deep sediments, hot springs), the new cells were originally called *archaeabacteria* to indicate that they might be the oldest of the three cell types [772–774]. Today, we know from metagenomic analysis that cells of this third type are abundant in all environments [775], and the three distinct realms of cellular life are now called *Archaea*, *Bacteria*, and *Eucarya* [776]. It is generally agreed (as we will discuss next) that the two types of prokaryotic cells preceded eukaryotic cells in the history of life on Earth. But there is no clear reason to conclude that one type of prokaryote evolved earlier than the other—or even that these are the only prokaryotic cell types that have ever existed.

Further study of *Archaea* confirmed the conclusions from ribosomal RNA analysis. This group differed from bacteria in a number of basic cellular features, including the structure of cell walls and membranes as well as the processes of replication, transcription, and translation. Certain complex biochemical processes seemed to be taxonomically limited to *Archaea*, such as methane production [773], and others were limited to bacteria, such as photosynthesis [777]. Nonetheless, genome sequence analysis provided evidence of horizontal DNA exchange between bacteria and *Archaea* living in similar environments [778, 779]. Intriguingly, it turned out that eukaryotic cells resemble archaeal cells in certain features (DNA compaction, replication, transcription, and translation) while resembling bacterial cells in others (membranes and metabolic pathways) **[769]** [779–782]. The greater similarities of basic genome expression processes in *Archaea* and *Eucarya* have led to multiple hypotheses about an *Archaea* as the progenitor of the eukaryotic cell, or at least of its nucleus [783–787].

The discovery of a third extant cell type has had profound consequences for thinking about the earliest stages of cell evolution on
Earth, for which there would be virtually no fossil record. If there are
now three surviving cell types rather than two, as previously believed,
what prevents us from hypothesizing that, at the earliest stages of
evolution, there once were additional cell types that have gone
extinct? Some scientists are beginning to search for evidence of these
earlier cell types by looking for taxonomic patterns indicative of horizontal transfers of particular DNA segments into a restricted subset
of both bacteria and archaea from extinct lineages [788, 789]. Finding
examples of such horizontal transfer would suggest that there were
older cell type DNA donors, and the nature of the horizontally transferred DNA would provide clues about the metabolic characteristics
of these now-extinct cells.

Symbiogenesis and the Origin of Eukaryotic Cells

Ever since extensive microscopic examination of prokaryotic and
eukaryotic cells began in the 19th Century, scientists have wondered
about the origins of the differences between the two cell types. Moreover, the use of the light microscope revealed that many organisms
were actually chimeras of two or more component organisms, often
involving microbes visible only by microscopy. As early as 1867, the
symbiotic nature of lichens (combining fungi and photosynthetic
algae) was described by Schwendener [790]. In 1888, the pioneering
Dutch microbiologist Martinus W. Beijerinck, who founded the Delft
school of environmental microbiology, succeeded in culturing live
bacteria from leguminous plant root nodules (http://www.asm.org/
ccLibraryFiles/Filename/0000000242/).

The importance of symbiotic associations in generating evolutionary novelty—and the related idea that eukaryotic cell organelles
are actually prokaryotic endosymbionts—were proposed repeatedly
in the late 19th and early 20th Centuries [791], most extensively by
Merezhovsky (1909) [792] and Kozo-Polyansky (1924) [793] in Russia, where they named the process *symbiogenesis*, and by Walin
(1927) in the United States, who used the comparable term *symbionticism* [794].

Most modern biologists ignore the importance and ubiquity of symbiotic associations, focusing instead on a small number of well-studied and domesticated *model organisms*. They operate on a strictly vertical model of genome transmission and generalize from laboratory results to nature as a whole. How misleading such a restricted focus can be is exemplified by the mid-to-late 20th Century discovery that so-called *plasmagenes* determining mating compatibilities in organisms as diverse as ciliate protozoa and insects are actually endosymbiotic bacteria [20, 795–801].

Cells living together through metabolic mutualism are everywhere in nature. When microorganisms are cultured for the ability to carry out specific metabolic tasks, the resulting isolate is often a *consortium* composed of different species (frequently both bacteria and archaea) that carry out complementary biochemical reactions [802–806]. In many cases, these metabolic consortia cannot be separated into their component microbes [807, 808]. Similar mutualistic dependencies exist with many larger organisms, such as orchids and cycads that depend on symbiotic fungi (*mycorrhiza*) to replace the normal functions of root systems [809–811].

Many invertebrates, such as termites, are metabolically dependent on their microbial flora, and some endosymbionts provide protection from predators or pathogens [812–815]. The gonads appear to be particular targets for endosymbiosis in invertebrates [816, 817]. There exist many specialized symbioses, where insects or other animals culture bacteria in special cells (*bacteriocytes*) and organs analogous to legume root nodules [818–820]. The light organ of the squid, which harbors bioluminescent bacteria at ultra-high concentrations, is a well-studied example [821, 822]. In these specialized symbioses, it is common to find that growth of the entire organism in the presence of antibiotics (which affect only bacterial cells) prevents reproduction of the host as well as the endosymbionts [823, 824]. In some cases, the progress of larval development depends on chemicals produced by the specially cultivated bacterial symbionts [825].

In any of these surprisingly widespread examples of essential inter-kingdom mutualism, we are unable to assert rigorously that the viable organism is composed uniquely of eukaryotic, plant or animal cells. It seems that we need to think of *organism* as a term that has a much broader community-based or systemic meaning than the significance given by traditional perspectives based on the idea that each organism has its own separate, vertically inherited genome [827].

The notion that eukaryotic cells originated by endosymbiosis has been firmly established by molecular analysis of two organelles that contain their own DNA and ribosomes: the mitochondrion and the chloroplast [828]. DNA and ribosomal RNA sequencing demonstrate that the mitochondrion is a descendant of an alphaproteobacterium and that the chloroplast descended from a photosynthetic cyanobacterium [829–832]. Thus, the capacity of eukaryotic cells to carry out oxidative metabolism and photosynthesis is clearly the result of symbiogenesis. In the case of the chloroplast (or *plastid*, as it is more widely known), the endosymbiotic event can be placed with reliability at the origin of photosynthetic algae and plants [833–837]. The timing of the origin of aerobic eukaryotes by alphaproteobacterial endosymbiosis is more difficult. Controversy exists over whether microbial eukaryotes that lack mitochondria don't have them because they evolved prior to the alphaproteobacterial endosymbiosis or, alternatively, because their ancestors lost a mitochondrion formerly possessed by the ancestral eukaryotic cell [838, 839]. Many eukaryotic cells possess degenerate forms of mitochondria, and a forceful school of thought argues that the root of all eukaryotes did indeed include the alphaproteobacterial endosymbiosis [840, 841].

Mitochondria and chloroplasts are not identical in all eukaryotes, and they show signs of major genome restructuring events over the course of evolution [842]. Plants tend to have much larger mitochondrial genomes than animals (100 to >700 kb compared to 6 to 20 kb), although they do not have proportionately greater coding capacity [832, 843–845]. Because the nuclear chromosomes encode the large majority of proteins that carry out mitochondrial and chloroplast biochemistry, it is clear that most of the genetic information in the original alphaproteobacterial and cyanobacterial endosymbionts was horizontally transferred intracellularly to the host nuclear genome [846–848]. The analysis of mitochondrial sequences inserted in nuclear genomes indicates that this intracellular horizontal transfer has been a recurring process.

Certain flagellated protozoa that include the major human pathogens *Trypanosoma* and *Leishmania* (causing sleeping sickness, Chagas' disease, and leishmaniosis) belong to the order *Kinetoplastida*.

This group received its name because organisms in it have very large exotic mitochondrial genomes that can account for up to 25% of the total cell DNA [849–851]. These *kinetoplast* mitochondrial genomes are composed of interlocking DNA rings (a *maxicircle* and multiple *minicircles*). Expression of these complex kinetoplast genomes is characterized by a process of posttranscriptional "editing" of mitochondrial RNA transcribed from the maxicircle DNA using *guide RNAs* encoded by the minicircle DNA rings **[80]** [852–855]. Why this elaborate mitochondrial genome structure and expression pattern evolved remains a mystery.

Among photosynthetic eukaryotes, the processes of secondary and tertiary endosymbiotic plastid evolution and intracellular horizontal DNA transfer between separate genome compartments have generated several major lineages [841, 856–858]. For example, the photosynthetic members of the phylum *Euglenophyta* are intensively studied for their circadian rhythms and tactic behaviors [859–863]. These algae diverged from a primitive group of freshwater flagellated protozoa when the original progenitor phagocytosed (engulfed) a green alga and integrated it as a plastid organelle [864, 865]. The euglenid cell thus contains four separate genomic compartments: nucleus, mitochondrion, a prokaryote-descended chloroplast, and a *nucleomorph* compartment descended from the algal nucleus [866, 867]. Genomics has established that both compartments in the alga-descended plastid have contributed DNA to the cell's nuclear genome.

The less well-known phylum *Chlorarachniophyta* resulted from a parallel secondary green algal endosymbiosis. Symbiogenesis of the equally obscure but ecologically important chromalveolate group of phyla, which includes ciliates, dinoflagellates, apicomplexans, brown algae, diatoms, and haptophytes, arose from secondary and tertiary endosymbiotic events involving photosynthetic red algae [856, 868]. While unfamiliar to anyone except specialists, many of these groups include major agents of global photosynthesis (such as diatoms [869–871]) and important human pathogens. Both *Toxoplasma gondii*, the toxoplasmosis pathogen, and *Plasmodium falciparum*, the malaria pathogen, are apicomplexan parasites that have lost the ability to carry out photosynthesis but still retain plastids. In all cases that have been examined, significant plastid-nucleus DNA transfer has occurred [872]. In addition to these symbiogenetic events at the base

of major phylogenetic groups, there are countless opportunistic symbioses, such as those where animals incorporate photosynthetic cells
and live off the products they produce using solar energy [873].

There is widespread conviction that the eukaryotic cell evolved
by a merger of two or more prokaryotic cells. This conviction has several sources. One arises from considering the metabolic requirements
for growth on a planet with increasing O_2 in the atmosphere as
cyanobacterial photosynthesis developed [874, 875]. Another source
of symbiogenetic thinking about eukaryotic origins involves the
apparent phylogenetic dichotomy in functional components: the
nucleic acid-based processes of DNA replication, transcription, and
translation resemble those in *Archaea* more than those in *Bacteria*,
while eukaryote membrane structure and metabolic pathways resemble *Bacteria* more than *Archaea*. Furthermore, the nucleus resembles an endosymbiotic prokaryotic cell enclosed in a distinct
membrane-bound intracellular compartment. This symbiogenetic
view typically sees the eukaryotic nuclear systems and translation
apparatus as being of archaeal descent, with the mitochondrion, cytoplasmic membrane, and perhaps other organelles derived from one
or more bacterial ancestors.

Examination of eukaryotic genome sequences and the relationships of distinct coding regions to one or the other prokaryotic
domain sequences also have been cited as evidence of the merged
nature of the original eukaryotic cells [876, 877]. Bioinformatic analysis indicates that the most ancient sequences found in living cells are
prokaryotic [878]. It is also possible, although rarely mentioned, that
one or more now-extinct cell lineages participated in the evolution of
the eukaryotic cell. There are many uncertainties about how the
extant lineages of eukaryotic cells may have evolved, but no serious
alternative to a symbiotic merger is proposed today to account for the
mixture of archaeal- and bacterial-like features in eukaryotic cells,
whose origin definitely ranks as one of the most important events in
evolutionary history.

Since the 1960s, Lynn Margulis, an irrepressible publisher of articles and books on this theme, has championed the importance of symbiogenesis as a primary source of evolutionary inventions **[688]** [879].
Margulis has applied the idea of endosymbiotic origins to include key
eukaryotic organelles—in particular, the microtubule-based cilia and

mitotic spindle **[152]** [880–882]. These organelles (and their variations in sensory organs and other tissues) share a characteristic "9 + 2" microtubule organization that is found throughout the eukaryotic realm. Margulis claims that cilia (or *undulipodia*, literally, "waving feet") descended from spirochete bacteria attached as motility commensals to a eukaryotic ancestor prior to the advent of mitosis [883]. Exactly these kinds of spirochete-protist associations are readily visible today in organisms from diverse ecosystems, such as the cockroach intestine. The chief difficulty in establishing the prokaryotic origin of undulipodia, centrosomes, or other organelles is that they do not have their own DNA or ribosomes, which would permit a solid phylogenetic identification. The only genomics-based method to confirm Margulis' hypothesis is to establish multiple clear relationships between the sequences of proteins from the organelle and a putative prokaryotic ancestor. The one detailed analysis carried out to date has failed to find prokaryotic homologies with the corresponding eukaryotic proteins [884].

Are Some Animals Hybrid Organisms with Hybrid Genomes?

Another highly controversial (but not illogical) symbiogenic evolutionary proposal has been put forward by Donald Williamson to explain the developmental history of invertebrates that display markedly different larval and adult stages, such as caterpillars and butterflies. The *larval transfer* or *hybridogenesis* proposal is that these organisms have combined two genomes in one dimorphic organism, such that one genome directs larval development and the other directs adult development [885–887]. This proposal solves the problem of why animals with clearly related adult forms differ discontinuously in their larval forms. But it has met heated opposition from mainstream developmental biologists. Much of the evidence Williamson cites in support of his ideas has not been published, and he does not propose clear molecular criteria to validate his proposal. So it is uncertain whether we will see another assumption overturned about eukaryotic evolution based on strictly vertical inheritance. Nonetheless, Williamson's idea merits mention as illustrating the testable new ideas that we are free to explore in 21st Century evolutionary theory.

Our understanding of how powerful an evolutionary force cell fusion has been is in its infancy. Although the ability of cells to merge and combine their genomes is a fundamental property of eukaryotic cells, we often take this capability for granted. But we should devote more attention to it. One has only to consider, for example, that cell fusion is the essential basis of all sexual reproduction to appreciate the central role it has played in evolution [888, 889]. Every fertilization event is both a symbiosis (cell merger) and a horizontal transfer of genomic material. This way of looking at sexual reproduction makes both symbiogenesis and horizontal transfer central to genome reproduction in complex eukaryotes. When the process is perturbed, as it is in interspecific matings, clear genomic and epigenomic consequences lead to novel forms (see Tables II.8 and II.10).

From a theoretical perspective, it is also important to remember that fertilization events involve the combination of hereditary elements that are not just composed of DNA [18]. It has become clear, for example, that the mammalian sperm *centrosome* (or *centriole*) plays a key role in zygote development [17, 18, 890]. The centrosome is the microtubule-organizing center of animal cells; it lacks DNA but displays a kind of self-templated inheritance [152] [11]. Largely because of our overwhelming focus on the genome, we are ignorant of how many other inherited organelles besides the centrosome are involved in normal fertilizations and how their reproduction is brought under coordinate control. Considering that the establishment of symbiosis is a less regular form of cell merger than fertilization, it should be obvious that we know very little about how two different cell cycles integrate so that all components of the symbiotic association can complete the duplication process coordinately. Without such integration, we would expect a newly established symbiotic association to break down, either because the symbiont does not keep pace with the host cell or because it outgrows the host.

We are currently learning about the plant-bacteria chemical dialogue and regulatory circuits that help establish nitrogen-fixing root nodules [891–895] and about the squid-bacteria collaboration to form a light organ [821]. In the root nodule case, it is interesting to note the role played by bacterial transport systems that are used in both DNA transfer and protein injection during pathogenesis [896]. Over 20 years ago, there was an intriguing report of synchronized sexual cycles involving a dinoflagellate host and its algal symbiont, but unfortunately the coordinating processes in this potential experimental system have not been further investigated [897]. A recent report documents a similar coordination between the circadian rhythms of a malaria parasite (*Plasmodium chabaudi*) and its rodent host [898]. This case may receive greater attention in the future because of its significance as a model system for a major human disease.

How cells merge successfully to generate a living novelty is obviously a critical area for future research. It is to be hoped that this formerly marginal subject will move closer to the center of biological research. Many well-identified endosymbiotic associations are now subject to experimental manipulation. So we can expect the next decade or two to teach us important lessons about communication and coordination between cell control circuits that were traditionally considered to be totally independent. It is difficult to imagine that such research will fail to deepen and extend our knowledge of symbiogenesis in evolution.

Natural Genetic Engineering and Evolutionary Genomic Innovation

The ability to obtain whole genome sequences has been of tremendous value in documenting the mechanistic basis of relatively recent evolutionary events. The time constraint on what can be inferred from genome sequence data exists because the distinguishing molecular markers of certain processes degrade as nucleotide sequences change over time. The time limit can be extended in the case of protein-coding regions because the amino acid sequences are statistically easier to follow over long periods.[1] We have already discussed the role of natural genetic engineering functions, such as DNA transposition and retrotransduction, in the exon-shuffling events that carry out domain rearrangements in protein evolution. Let us now turn our attention to other kinds of genomic innovation and see what has been established about the potential roles of natural genetic engineering functions in generating those novelties.

1   The statististics work out this way because there are 20 possible amino acids at each position of a protein chain but only four possible bases at each position of a nucleic acid chain.

Novel Exons

One of the principal observations in the Evo-Devo field that studies the evolution of development networks has been the emergence of novel protein domains at critical stages where there have been morphological advances. The focus on domains has been so intense that developmental biologists refer generically to many developmental regulatory circuit families by their characteristic domains, such as *POU domain circuits* or *MADS-box networks* in animals and plants [899, 900]. The best-known example of such a critical domain is probably the *homeodomain*, a DNA-binding region common to many different transcriptional regulatory proteins in animal development [901].

To have new subprotein domains arise in the course of evolution, a process is needed for generating novel exons that can encode extended polypeptide structures to be incorporated into proteins in combination with other exons. Exon generation cannot occur efficiently by the gradual accumulation of single amino acid changes in existing protein chains because the probability of losing the original functionality is too high and of gaining a new functionality too low. A more rapid, facultative process is needed—and has in fact been discovered.

Exonization is the name given to the appearance of a novel protein coding region that can be spliced into an mRNA molecule. Exonization occurs when a transposon or retrotransposon inserts into an existing genetic locus [902–905]. Mobile elements contain potential coding sequences (in all reading frames) and also potential splice sites. Consequently, mobile element insertion into a transcribed but untranslated region or into an intron (much more rarely into an exon) changes the arrangement of available exons, splice sites, and introns in the primary transcript. If the splice sites within the mobile element are utilized, and if the resulting RNA sequence contains a continuous in-frame coding sequence, a protein containing a novel polypeptide segment will be translated [906].

Exonization has been documented for all classes of mobile genetic elements. Initially, this process was missed because repeats were eliminated from coding sequences to be analyzed by applying a program called Repeat Masker. This was done because it was assumed that all protein coding sequences had to be unique. However, when Nekrutenko and Li analyzed protein coding sequences without previous Repeat Masker screening in 2001, they discovered that a significant fraction (~5%) contained regions corresponding to common interspersed mobile elements [907]. Mobile element insertions are abundant in the introns of mouse and human genomes. A recent survey of these two genomes found that exonization is more frequent in humans than in mice [908]. About 3,500 of the estimated 26,000 human "genes" contain exons originating from mobile elements, whereas only 1,200 out of a similar number of mouse "genes" have them.

The greater frequency of exonization in humans as compared to mice is at least partly attributable to a high degree of exonization by the primate-specific *Alu* SINE elements, which are not found in rodents or other mammalian orders [909]. Thus, an exonized *Alu* creates a primate-specific coding element. A great deal of current speculation centers on the role of *Alu* exonization in the apparently rapid evolution of hominids and other primates.

In addition to exonization, mobile element-encoded proteins can be a source of coding sequences adapted for novel functions. Simple examples include the role of transposases evolved for specialized chromosome breakage in yeast mating-type switching and immunoglobulin VDJ joining. A potentially more complex example depends on extending the evidence that retroviral proteins evolved into essential imprinted placental functions [910–915]. In other words, it may be the case that LTR retrotransposons played a role in the origins of the placenta, a key invention in mammalian evolution [916]. Not only retrotransposon coding sequences but also regulatory modules seem to have participated in fundamental mammalian evolution. Parallels have been observed between the epigenetic control of preimplantation embryonic development, of placental growth, and of mobile elements [917–919].

Novel Introns and Alternative Splicing

In addition to generating novel exons, insertion of mobile elements can create new introns and thus split existing exons into smaller ones. The mobile homing endonuclease-dependent group I self-splicing introns and retrosplicing group II introns were discussed in Part II of this book. Susan Wessler first described the capacity for *intronization* of other classes of mobile elements in 1987 with McClintock's *Ds* transposon, which can be spliced out of a transcript to restore function to an interrupted genetic locus [920].

Wessler also discovered that *Ds* insertions are subject to alternative splicing, using different splice sites within the transposon [921]. This result indicated that mobile element insertions may be an important source of alternative splice sites available for expression of distinct protein products in different tissues or under different conditions. The exonized mobile elements found in mammalian and other genomes are known to be subject to alternative splicing, an important means of extending coding diversity [922–925].

Regulating the Speed of Transcription

An intriguing observation is that LINE elements can slow down the rate of transcription when inserted into an intron. Given their abundance in mammalian genomes, especially our own, LINE elements have been called *modulators* of genome expression [926–928].

Novel CRMs and Cis-Regulatory Networks

Barbara McClintock called her mobile elements "controlling elements" because they altered the regulation of a genetic locus where they were inserted [4, 233, 929]. The first DNA transposons in bacteria were identified by their ability to terminate or initiate transcription [930–932]. Early studies with LTR retrotransposons in yeast demonstrated their ability to stimulate expression by contributing strong enhancer elements to different loci [492], and retroviral oncogenesis by insertional mutations resulted from altering transcriptional regulation [488, 933–935].

Since the early period of mobile element research in the 1970s and 1980s, growing experimental and genomic evidence has indicated that insertion of mobile elements contributed to the formation of many *cis*-regulatory modules (CRMs) [195–197, 208, 209, 211, 936]. Because each species and genus has its own constellation of mobile elements, it is to be expected that *cis*-regulatory changes that occur in this way will differ from one lineage to another [210]. An intriguing illustration of this point is evident in rodents and primates, where different LTR retrotransposons participated in creating distinct but functionally equivalent transcriptional control sites for expressing the same apoptosis protein [937].

The potential for dispersed repeats to group unlinked loci so that they share common regulatory signals was first demonstrated by McClintock in 1956 [233]. The discovery of repetitive DNA by Roy Britten and his colleagues in the late 1960s **[377]** made it apparent that dispersed repeats could provide the physical basis for the construction of regulatory networks in the genome [75]. In a few cases, sequence information has made it possible to establish mobile elements as the agents responsible for initiating transcription at suites of distinct genetic loci. One example involves the vertebrate REST (Repressor Element Silencing Transcription) regulatory factor that participates in many cell proliferation and physiological response circuits. Evolution of REST-responsive *cis*-regulatory sites has involved repeated insertion of LINE elements [938–940]. Other examples include a c-Myc regulated network in our own genomes [942] and a family of LTR retrotransposons that initiate transcription at various genomic sites during oogenesis and early embryonic cell divisions in mice [201, 941].

Wide-scale examination of coordinately transcribed regions of the genome is becoming routine with modern DNA microchip technologies. However, it is not a simple matter to determine which regulatory signals are responsible for controlling the observed transcriptional patterns. When that has been accomplished through bioinformatics and direct experimentation, we will have a better idea of how many regulatory suites evolved by mobile element activity.

Epigenetic Imprinting Sites and Chromatin Domains

Epigenetic chromatin formatting provides a higher level of expression control extending over large genome regions. At the same time, RNA-directed chromatin formatting serves as a generic silencing routine to hold the activity of mobile genetic elements in check. Because chromatin formatting can spread along a chromosome until it encounters an insulator signal, the insertion of a mobile element at a new genomic location typically alters the nature of chromatin surrounding the target site. Such localized chromatin alteration accounts for the phenotypic effects of many mobile element mutations.

The connection between mobile elements and chromatin formatting suggests that chromatin domains can be established during evolution by the insertion of transposons and retrotransposons [943–945]. It is certainly the case that the majority of heterochromatic regions in eukaryotic genomes (besides specialized centromere and telomere domains) are rich in mobile elements [315, 316, 946], and mobile elements play a key role in centromere evolution as well [947–950]. Some of the strongest evidence to date for the role of mobile elements in establishing specific epigenetic controls comes from the study of imprinted loci in both plants and animals. Imprinting signals are often repeats near the transcription start site, and in many cases, the repeats are clearly derived from SINEs or other mobile elements [919, 951–955]. In addition, we know that the *gypsy* retrovirus in *Drosophila* carries a strong insulator element that can alter the extent of chromatin domains [499]. Thus, the potential for epigenetic formatting by mobile elements has been well established, and it remains to define the role such formatting has played in the evolution of functionally important chromatin domains.

Novel Regulatory RNA Molecules

Molecular, cell, and developmental biologists are currently engaged in the enormous task of incorporating recent discoveries about regulatory RNA molecules into their models of how vital functions are controlled. The realization that so-called *noncoding* ncRNAs play such a pervasive role in controlling cell functions is only about 15 years old. Discoveries about the breadth and depth of the ncRNA control circuits continue to accelerate **[296, 300, 956, 957]**. It is a popular argument to assert that newly recognized ncRNA control regimes compensate for the disappointment in finding less protein-coding capacity than expected in human and other mammalian genomes and that ncRNA provides the predicted evolutionary complexity [958, 959].

Given the central role now recognized for ncRNA regulatory circuits, it is logical to ask what role natural genetic engineering may have played in the elaboration of these control regimes. In plants, many small ncRNA molecules have derived from miniature DNA transposons called *MITEs* [960, 961] or, in rice, from the same Pack-MULE elements involved in exon shuffling [962]. Plant molecular geneticists have even suggested that ncRNA regulation originally evolved from a more primitive defense regime against the activity of mobile elements [963].

Experimental evidence and genome analysis further indicate that taxonomically specific mobile elements are sources of ncRNA regulation. There are ancient and conserved ncRNAs that correspond to mobile elements that appeared early in mammalian evolution, but there are also marsupial-specific and primate-specific ncRNAs that correspond to marsupial-specific and primate-specific mobile genetic elements [964–968]. In the human genome, 55 functionally characterized and 85 uncharacterized miRNAs arose from transposons and retrotransposons [217]. In other words, the potential for evolving new RNA regulatory regimes depends, at least in part, on the available repertoire of dispersed mobile repeat sequences that can be used to make new ncRNAs.

Chromosome Rearrangements

Cytogeneticists have known for decades that speciation within particular taxonomic groups involves two complementary processes: chromosome structural rearrangements on a large scale and conservation of chromosome organization between the rearrangement sites or *breakpoints* [969–973]. This pattern has been confirmed by genome sequencing. Related organisms share long *syntenic* (literally, "on the same band") chromosome segments that contain the same sequences of genetic loci, indicating that they evolved from a common ancestral chromosome segment [974–977]. However, dozens or hundreds of syntenic regions in related genomes may be scrambled and assembled into markedly different chromosome arrangements.

Between the human and mouse genomes, for example, the initial mouse genome sequencing effort in 2002 recognized 342 syntenic segments larger than 300 kb in length distributed differently (but not randomly) between the chromosome sets of the two species [978]. The scrambling in order and orientation is evident, but there are tendencies for segments corresponding to a particular human chromosome to be located on a small subset of the 20 mouse chromosomes. The conservation of syntenic segments is interpreted as the tendency to maintain clusters of genetic loci that are coregulated in some way, possibly in part by epigenetic chromatin domains. The restrictions on chromosome placement of the scrambled segments may reflect a still higher level of regulation, perhaps by subnuclear localization.

A single eukaryotic genome often contains evidence for chromosome rearrangements in the evolutionary process. As you will see shortly, many eukaryotic genomes evolved from whole genome duplication (WGD) events in which all the chromosomes doubled in number. The traces of these WGD events are the presence of numerous localized *segmental duplications* in the evolved diploid genomes that have lost most of the doubled chromosomal material [979–982]. These segmental duplications are rearranged syntenic regions distributed through the genome—sometimes on the same chromosome, sometimes on different chromosomes, sometimes in the same orientation, sometimes in opposite orientations. The poorly understood process leading from a completely duplicated genome to one that contains only segmental duplications is called *diploidization* [983–985].

Clearly, in all the chromosome rearrangements seen either by syntenic segmental duplications within a genome or by comparisons of syntenic regions in related genomes, processes of chromosome breakage and resealing in novel combinations have been at work. In other words, chromosome rearrangements occurred that necessarily involved natural genetic engineering systems. The sites of these breakage and resealing events are called *evolutionary breakpoints*, and a growing effort is under way to analyze the nature of these breakpoints [986–989].

Certain sites that have been used repeatedly as breakpoints are called *fragile sites* or *fragile regions* [990–994]. It is possible to identify dispersed repeat elements with many of these recurrent breakpoints [995–997]. In *Drosophila*, some recurrent breakpoints are known to be DNA transposons [998–1002]. A recent comparison of the human and gibbon genomes found a high association of new and pre-existing mobile element insertions at evolutionary breakpoints [1003]. Fragile sites for chromosome rearrangement are also important in cancer [1004, 1005], and there has been an effort to correlate tumor and evolutionary breakpoints [1006].

The most common interpretation of recurrent breakpoint patterns is that dispersed repeats at different chromosome locations serve as regions for DS break repair by homologous recombination [418, 420]. However, it is also possible that non-homologous DNA rearrangement processes, such as those demonstrated experimentally for *Drosophila* DNA transposons, have been at work [462, 1007]. Whatever the final array of evolutionary chromosome rearrangement mechanisms turns out to be, it is becoming ever more clear that an intimate connection exists between rearrangement breakpoints, repetitive DNA, and mobile genetic elements.

Use and Reuse of Evolutionary Inventions

In the late 18th and early 19th Centuries, the main focus of evolutionary studies was fossil evidence for repeated changes in the diversity and nature of living organisms over the course of Earth history. Pioneers such as Georges Cuvier in France, the founder of comparative vertebrate anatomy, established two basic aspects of the evolutionary process that were then new to science:

- The extinction of many organisms found in older strata of the geological record, followed by their replacement with new organisms in more recent strata
- The frequent maintenance of morphological relationships between extinct and successor organisms [1008, 1009]

Although Cuvier himself did not recognize the full implications of these new scientific principles, they form the basis of our contemporary thinking about the succession and evolution of novel life-forms. Descent by modification and replacement of older life-forms with newer ones were the principles championed by the Darwins, Erasmus in the late 1700s, and his grandson Charles in the second half of the 1800s [1, 1010–1013].

Descent with modification provides the overall context for this book, whose main theme is to illustrate how many exciting facts we have learned about the processes that lead to evolutionary inventions. Analyzing the fossil record is somewhat outside the scope of this book, but the correlation of paleontological novelties and genome organization is a fascinating question addressed by the branch of science now called Evo-Devo, the study of the evolutionary basis of morphogenetic processes **[1014]** [1015–1018]. Evo-Devo attempts to integrate the results of molecular developmental biology, genomics, and paleontology. Molecular development tells us about the networks and component molecules that guide the morphogenesis of contemporary organisms. Genomics permits us to infer phylogenies of those morphogenetic circuits, and paleontology tells us about the historical record. As the three areas of study become more fully integrated, our confidence grows in the solidity of our understanding of the evolutionary process.

The greatest success of the Evo-Devo approach has been to demonstrate the surprising conservation of morphogenetic routines over long periods of time. The most outstanding single case is the discovery of the *homeobox* or *Hox* complex. It plays a central role in executing morphological differentiation along the anterior-posterior (AP, or head-to-tail) axis of all bilaterally symmetrical animals (*Bilateria*) **[1019–1021]** [1022–1025].

The Hox complex was initially discovered in *Drosophila*, where it was the site of mutations that led to so-called *homeotic* transformations of one body appendage into another (for example, an antenna into a leg) or of one body segment into another (for example, creating a four-winged fly with two middle thoracic segments instead of one) [1026]. Later, in the 1980s, it was unexpectedly found that vertebrates (and, subsequently, all other animals in the *Bilateria* group) have Hox complexes and use them in similar ways to *Drosophila* to execute differential morphogenesis along the AP body axis [1027–1030]. This was quite a surprise, because the processes of embryonic development in insect and vertebrates are completely different. Apparently, the Hox complex was so useful that it was maintained, while many other aspects of morphogenesis changed in fundamental ways.

The Hox complex is a higher-order genomic structure. Each Hox complex combines a series of eight to ten coding sequences for homeobox-domain (Hox protein) transcription factors interspersed with an array of transcriptional regulatory and chromatin formatting signals organized into complex cis-regulatory modules (CRMs) [1031–1033]. These CRMs guide the progressive expression of the different Hox proteins at different positions in the body, and the Hox proteins in turn regulate the correct expression patterns to generate the appropriate body structures. To apply the computational metaphor, the Hox complex can be compared to a specialized microchip dedicated to controlling morphogenesis from head to toe.

There has been a longstanding debate about the original evolution of the bilateral body plan and the Hox complex [1034–1039]. It is known from genome sequences that homologues of the individual Hox proteins appear in animals that do not belong within *Bilateria* (such as sea anemones, corals, jellyfish, and hydra) [1040]. So the Hox proteins themselves appear to have an older evolutionary origin in the *Metazoa* (multicellular animals). What is unique to *Bilateria* is the linear order and grouping of the coding sequences for the various Hox proteins. That order has been maintained throughout the evolution of bilaterally symmetrical animals [1033]. Thus, we may speculate that it was the assembly of the component coding and regulatory elements into a functioning higher-order Hox complex that was critical, not the evolution of the individual Hox proteins. In other words, natural genetic engineering to put together the Hox complex may have been a key process leading to *Bilateria* [1041].

The correlation between Hox complex evolution and the fossil record brings us to the two main paleontological events of animal diversification. In the Ediacaran period (635 to 542 million years ago (MYA)), fossils of the first soft-bodied macroscopic animals appeared [1042–1044]. During the famous Cambrian explosion (542 to 500 MYA), most of the Ediacaran fauna went extinct, and all the phyla of existing *Bilateria* abruptly appeared in the fossil record **[1045, 1046]** [1047–1050]. The oldest *Bilateria* fossil is dated to 555 MYA, in the late Ediacaran [1051]. This indicates that it took some time for the *Bilateria* to diversify and then take over when the early Ediacaran fauna disappeared. Clearly, many inventions besides the Hox complex were needed for the *Bilateria* radiation to occur.

What the Hox complex illustrates in a dramatic way is the principle that evolutionary inventions are retained and reused within the context of new organismal developmental programs. The developmental differences between *Drosophila* and mammals are enormous. *Drosophila* has a number of features characteristic of higher insect embryogenesis: a multinucleate syncytial embryo; compartmented multicellular embryos; metamorphosis into distinct larval, pupal, and adult stages; and adult appendages that develop in compressed form inside compartments called *imaginal disks* **[1052–1055]** [1056]. Mammals, on the contrary, like all vertebrates, develop by direct cell division from the initial zygote, undergo specific folding events (such as gastrulation and neural tube closure), and gradually enlarge new tissues and limbs by growth of primordial structures **[1057, 1058]**. Nonetheless, the Hox complex does a similar job in both insects and mammals despite their tremendous embryological differences **[1020]**.

The same principle of retention and reuse of evolutionary novelties is illustrated by the relatively small number of intercellular signaling circuits that operate in various *Bilateria* but that do their jobs integrated into quite distinct morphogenetic programs **[1021]** [1059–1062]. Even within a single organism, these signaling circuits are used repeatedly and work to develop distinct limb structures as well as nerves, muscles, intestines, and other tissues and organs. This reuse in very different morphogenetic contexts tells us that it is not the molecular machinery itself that determines form. Rather, proper form results from the control architecture that governs how Hox complexes, intercellular signaling circuits, and other essential evolutionary inventions are mobilized at each stage of multicellular development. At present, our understanding of basic principles governing this overall control architecture is severely limited, and it certainly deserves to be a prime subject of 21st Century research.

What Makes a Man Different from a Mouse?

In the film *A Day at the Races*, a straight-laced character aggressively asks Groucho Marx, "Are you a man or a mouse?" Groucho answers nonchalantly, "Put a piece of cheese on the floor, and you'll find out!" If only it were that simple to determine what makes two related organisms appear so different to our eyes in their outward characteristics and behaviors.

We have already discussed some of the genomic features that distinguish men and mice. The most outstanding genomic features are the scrambling of syntenic regions in the chromosome complement and the distinct repertoire of dispersed mobile repeat elements in the two species. Because we have seen how these dispersed repeats can be the sources of genomic novelties, it is worth looking at this difference in more detail. Mouse and human genomes share some basic mammalian mobile elements, such as DNA transposons and the MIR (Mammalian Interspersed Repeat) and LINE retrotransposons. However, the most numerically abundant repeats are the SINE retrotransposons, and they differ completely between the two genomes. The human genome contains about 1.1 million copies of the *Alu* SINE repeat element that is specific to primates and absent from rodents [382], while the mouse genome contains over 900,000 B1 and B2 SINE elements not found in the human genome [978]. In other words, at least 2 million retrotransposition events separate the human and mouse genomes.

What differences do these distinct repertoires of SINE elements make? We are still largely ignorant of how they influence the large-scale physical organization and regulatory architectures of the two genomes within the nucleus. Our ignorance notwithstanding, it is important to note that SINE elements contain promoter, enhancer, splicing, mRNA targeting, and translation stimulation signals that can affect genome expression, as well as sites for epigenetic marking [62, 203]. In addition, there is increasing evidence for the role of SINE transcripts in ncRNA regulation [965, 966]. Thus, it is almost certain that the SINE element differences will have a significant influence on how the two genomes function, especially how they are differentially expressed during development. The traditional view has been that related species differ in their repertoire of individual "genes." But a more contemporary Evo-Devo perspective is that much of morphological change in evolution occurs by modification of expression through alteration of enhancers and other transcriptional regulatory signals, as well as distinct patterns of epigenetic formatting [1063–1067].

Comparing mice and men, the "genes" stay largely the same, but their deployment differs. The bones, ligaments, muscles, skin, and other tissues are similar, but their morphogeneses and growth follow distinct patterns. In other words, humans and mice share most of their proteins, and the most obvious differences in morphology and metabolism can be attributed to distinct regulatory patterns in late embryonic and postnatal development. The distinct SINE repertoires in mice and men actually have contributed to distinct regulatory signals (about 5% of human promoter regions contain an *Alu* element) and also to the evolution of taxonomically specific proteins. About 5% to 10% of human and mouse proteins contain exonized SINE element sequences [907]. Those proteins could only have evolved in either the primate or rodent lineage because of the restricted distribution of the exonized SINEs [385]. It will be of great interest to learn how many of these taxonomically specific proteins actually do play a role in the distinct metabolic, morphological, or behavioral features that (like the response to cheese) distinguish a man from a mouse

Whole Genome Doubling at Critical Stages of Evolutionary Innovation and Divergence

One of the main differences between invertebrates and vertebrates is in the number of Hox complexes. Invertebrates have only one (including the *Chordata*, which share a dorsal nerve column with vertebrates), but there are four Hox complexes in mammals and most other vertebrates and eight in bony fish (*Teleosts*) [1025, 1040, 1068]. Salmon have 13 Hox complexes, indicating a further amplification event in this fish lineage [1069].

Where did these extra Hox complexes come from? The answer comes from noting that duplications, triplications, tetraplications, and even octaplications are not limited to Hox complexes. They also are observed for other important genetic loci, such as those encoding surface receptors and components of signal transduction circuits [1070]. The sources of such widespread amplifications are now seen to be the result of "whole genome duplications" (WGDs). In his prescient 1970 book *Evolution by Gene Duplication*, Susumu Ohno predicted two successive WGD events in the evolution of vertebrates [1071]. These two WGD events— the first preceding appearance of the first verte-brates, and the second preceding appearance of jawed vertebrates— produced the required four copies of the Hox complexes and other genome components [1072–1074].

Rigorously speaking, genomic evidence alone is insufficient to establish that a WGD has occurred. What is observed in cases where we believe that a WGD has occurred is the presence of a significant number of duplicated syntenic regions throughout the genome. Technically, this can only be termed a *large-scale duplication* (LSD) event, conceivably having arisen through a series of independent or even coordinated segmental duplications in different parts of the genome. Nonetheless, geneticists, genomicists, and evolutionists are confident that WGD is the correct explanation for these dispersed duplications, because we can actually observe WGD and its direct role in speciation in real time.

It is important to note that selection has never led to formation of a new species, as Darwin postulated. No matter how morphologically and behaviorally different they become, all dogs remain members of the same species, are capable of interbreeding with other dogs, and will revert in a few generations to a common feral dog phenotype if allowed to go wild. The way we make new species synthetically is by interspecific hybridization. The importance of interspecific hybridization has been mentioned by many early evolutionists, including Lamarck, and is documented in the scientific literature back at least to the 19th Century. The cereal plant *Triticale* was cre-ated in this way by crossing wheat (*Triticum*) and rye (*Secale*); it is currently cultivated in Europe and China **[1075]** [1076–1078]. The most important proponent of what he called *cataclysmic evolution* **[643]** was plant cytogeneticist and evolutionist G. Ledyard Stebbins [1079, 1080]. He studied plants of the mustard family (*Brassica*) and demonstrated that interspecific hybridization led to the formation of new species (such as mustard greens or rapeseed) with chromosome numbers equal to the sum of both parental genomes (http://www.answers.com/topic/g-ledyard-stebbins). Similar experiments were performed in the 1920s by Nikolai Vavilov and his student Georgy Karpechenko in the Soviet Union before Lysenkoism decimated Soviet genetics (M. Golubovsky, personal communication) [1081].

Chromosome doubling occurs with interspecific hybridization because hybrids without WGD cannot go through meiosis and therefore are sterile. Sometimes, as with crosses to produce new *Triticale* hybrids, the parental genomes are artificially doubled before hybridization by treating the plants with colchicine, a microtubule inhibitor that blocks chromosome separation in meiosis. But the formation of diploid gametes with doubled genomes is common (occurring in about 1% of all mouse fertilizations [1082]). As a consequence, hybrids with doubled genomes can form naturally at reasonably high frequency. In *Arabidopsis* interspecific hybrids, about 25% underwent spontaneous chromosome doubling and were fertile [1083]. It is clear from genomic analysis what Stebbins and other plant breeders have long known—that hybridization and genome duplication have been the sources of new plant species. Similar speciation events resulting from interspecific hybridization have been observed in butterflies, moths, and other animals [1084–1087].

In a letter to his friend, the botanist J.D. Hooker, on July 22, 1879, Darwin called the rapid diversification of flowering plants (angiosperms) in the fossil record of the Lower Cretaceous "an abominable mystery" [1088]. It is now clear that a series of WGD events played repeated roles in angiosperm evolution [1089–1092]. Rapid plant evolution was "abominable" to Darwin because formation of interspecific hybrids and genome doubling are the kinds of sudden, genome-wide changes affecting multiple characters that he explicitly excluded from his gradualist, uniformitarian thinking.

In addition to flowering plants and vertebrates, we know that WGDs have played a role in the evolution of fungi and protozoa [1093–1096]. It is likely that even more instances of WGD at major evolutionary junctures will be documented as more eukaryotic genomes are sequenced. One of the most significant features of WGD events is that they produce two copies of the dispersed genome regions that encode complex networks [1097, 1098]. Having an extra copy of the entire network means that no functionality is lost if one copy of the network is modified to change its inputs, outputs, and/or internal operation. The fact that intracellular signaling networks, for example, have been adapted to many different cellular and developmental functions indicates that, in the course of evolution, they have been duplicated and modified to meet new adaptive needs

[1099, 1100]. It makes sense to believe that this kind of whole-network adaptation is significantly easier to execute in an organism that has a recently duplicated genome than in one where every network is unique and fulfills an important functional requirement. There is a striking and highly significant connection between what we have learned about the molecular basis of genome change and the role of WGDs in evolutionary history. Formation of interspecific hybrids and changes in ploidy are well-documented *genome shocks* that lead to the disruption of epigenetic control on mobile genetic elements in both plants and animals. This accounts for the numerous cases where these elements have played important roles in evolutionary change (see Table III.2). The potential for life-history-based control over the occurrence of hereditary variation is one of the most trenchant and fundamental differences between the 21st Century view of genome change resulting from a constellation of regulated cell functions (natural genetic engineering) and the traditional view that genome change results from random and accidental events. Part IV of this book explores some of the deeper conceptual implications of this difference.

Table III.2 Natural Genetic Engineering Rearrangements Documented in the Evolution of Sequenced Genomes (References Appear in the Online Version)

Pack-MULE transposons mediating coding sequence duplications and exon shuffling in rice

Exon shuffling by a CACTA transposon in beans (*glycine max*)

Exon shuffling and amplification by helitrons in maize

Exon origination in coffee and *Arabidopsis* from transposable elements

The *Hobo* transposon involved in endemic inversions in natural *Drosophila* populations

Gross chromosome rearrangements mediated by transposable elements in *Drosophila melanogaster*; the data include natural populations

Generation of a widespread *Drosophila buzzatii* inversion by a transposable element; two natural hotspots and multiple other rearrangements in the *Drosophila buzzatii* genome induced by the *Gallileo* transposon

Penelope and *Ulysses* retroelements involved in *Drosophila virilis* chromosome rearrangements at natural breakpoints

Chromosome rearrangements involving two transposons

Hotspots in transposon-generated chromosome rearrangements

Table III.2   Natural Genetic Engineering Rearrangements Documented in the Evolution of Sequenced Genomes (References Appear in the Online Version)

Abundance and recent occurrence of segmental duplications in the human genome

Segmental duplications found at syntenic region breakpoints in human and mouse genomes

Role of transposable elements as chromosome rearrangement catalysts

Richness of transposable elements in *Drosophila* pericentric heterochromatin

Novel transposable element insertions found near loci encoding insecticide-metabolizing enzymes in *Drosophila*

Segmental duplication associated with a chromosome inversion in malaria mosquito vector

Dispersed LINE and SINE repeats in the human genome as substrates for ectopic homologous recombination

Coincidence of primate syntenic breakpoints with presence of transposable elements

LINE-1 elements associated with deletions in human genome variation

DS breaks associated with repetitive DNA in yeast

Many inversions associated with L1 repeats

Syntenic breakpoints between human and gibbon genomes show new insertions of gibbon-specific repeats and mosaic structures involving segmental duplications, LINE, SINE, and LTR elements

Chromosome rearrangements by Ty element recombination in a wild strain of yeast used for wine fermentation

Evolutionary breakpoints in wallaby genome associated with SINEs, LINEs, and endogenous retroviruses

P element insertions next to heat shock promoters in wild *Drosophila*

Transposable element clusters at syntenic breakpoints in three *Entamoebae* species

Reviewing What the DNA Record Reveals about Cell Activities over Evolutionary Time

In Part II, you learned about the active cell processes that lead to hereditary changes in the genome. In this third section of the book, we've examined the genome DNA record to see what cell functions may have played a role in change over evolutionary time. The examination revealed a number of major surprises, starting with the discovery of widespread horizontal transfer of DNA between prokaryotes and the growing evidence of horizontal transfer in eukaryotes. Transfers occur in real time within and between the three basic kingdoms of life. They occur by uptake of naked DNA, direct cell-to-cell DNA transfer, and transduction by viral or virus-like particles. With so much evidence of horizontal exchange, we can no longer predicate our views of evolutionary change uniquely on the basis of vertical transmission of genomic information.

Next, using the vast and continually expanding sequence databases, we came to realize that proteins share functional segments, or domains, far more widely than they share overall structure. This means that proteins evolve by accumulating and rearranging polypeptide domains rather than by undergoing a series of individual amino acid changes. The underlying genomic processes are not stochastic, localized point mutations, but rather exchanges of DNA segments encoding the polypeptides that comprise these domains. Through laboratory experiments and genome analysis, it has been possible to identify a number of the molecular processes that carry out domain shuffling. Typically, these involve the action of transposons or retrotransposons.

The attempt to place taxonomy on a solid molecular basis turned up yet another surprise. Rather than two basic cell types, living organisms can be divided into three equally separate types: *Bacteria*, *Archaea*, and *Eucarya*. The recognition of an unexpected cell type (*Archaea*) reminds us that the history of life may have included additional kinds of cells, now extinct but which contributed to the surviving cells we currently study. Classification of DNA from mitochondria and chloroplasts showed that symbiogenic cell fusions were key events in the formation of eukaryotic cells. Most (if not all) eukaryotes descended from the original mitochondrial merger, and all photosynthetic eukaryotes descended from the original chloroplast merger. There is growing recognition of how numerous secondary and tertiary symbiogenic events have occurred to originate otherwise-distant photosynthetic lineages. We are rapidly accumulating evidence for past and current symbiotic relationships within and between all the major kingdoms of life. The full role of symbiogenesis in evolution has yet to be appreciated or documented. Any comprehensive theory of evolution has to include symbiotic mergers in the patterns of genome inheritance.

Examination of genome sequences at a detailed level has uncovered a continual process of innovation of the many different components needed for DNA-based information storage and genome formatting as an information organelle. Coding sequences, regulatory motifs, signals for DNA compaction and epigenetic control, and repetitive DNA elements all show taxonomically specific patterns of origination. In many cases, new coding sequences (exons) or regulatory motifs appear as the consequences of mobile genetic elements moving to new locations.

At a higher level, analysis of chromosome structure and organization reveals a dynamic history of structural rearrangements combined with a surprising degree of conservation of syntenic regions. Within any given genome, syntenic regions are completely or partially duplicated, indicating processes of partial or whole genome duplication (WGD). In organisms displaying extensive syntenic duplications, we have been able to infer a history of WGD events because this same process can be observed in real time, often as a consequence of interspecific hybridization. The evidence shows that interspecific hybridization and WGD are key events in the formation of synthetic species, something that has not been achieved by selection. WGD events have been documented in widely divergent taxonomic groups, including protozoa, yeasts, vertebrates, and flowering plants. Thus, we have to include hybridization and genome doublings in our catalog of exceptions to normal vertical inheritance as major triggers of evolutionary change.

Examining genomes and deducing what kinds of hereditary changes occurred coincidentally with major transition points in evolution (when new kinds of organisms appeared having novel capabilities) lead to a clear conclusion: Rapid events involving non-canonical modes of inheritance have introduced major changes to genome structure and function throughout evolutionary history. The DNA record definitely does not support the slow accumulation of random gradual changes transmitted by restricted patterns of vertical descent.

The final part of this book discusses how we can integrate what we have learned about cell information processing (Part I), genome function as a RW memory system (Part II), and the DNA record of cell and genome changes in the course of evolution (Part III). The aim is to formulate a coherent (but not restrictive) contemporary view of organic evolution that will fit into the real-life history of organisms on a continually changing planet.

A New Conceptual Basis for Evolutionary Research in the 21st Century

In this part:

- A systems approach to generating functional novelties
- Reorganizing established functions to generate novelty
- Generation of novel components
- Retention, duplication, and diversification of evolutionary inventions
- The implications of targeting genome restructuring
- Can genomic changes be linked to ecological disruptions?
- What might a 21st Century theory of evolution look like?
- Where does evolution fit in 21st Century science?

General discussions of evolution, especially in the context of the "Intelligent Design" controversy, suffer from an unfortunate conflation in the minds of the lay public (and also of scientists) of three distinct questions:

- The origin of life
- The evidentiary basis for an evolutionary process
- The nature of evolutionary change

Almost universally, the term *Darwinism* is assumed to be synonymous with a scientific approach that has provided satisfactory answers to all three questions. It is to be hoped that, by now, you realize that these three questions are individually complex and that two of them are quite far from having coherent scientific explanations.

We have little solid science on the origin of life, in large part because there is virtually no physical record, but also because we still have gaps in our understanding of what constitute the fundamental principles of life. As to the actual nature of evolutionary change processes, you have seen in Parts II and III that cytogenetic observations, laboratory experiments, and, above all, molecular evidence about genome sequence changes tell us that the simplifying assumptions made in the 19th and early 20th Centuries are plainly wrong. They fail to account for the variety of cellular and genomic events we now know to have occurred. It should be emphasized that many change events have been quite rapid and have involved the whole genome—notably, symbiosis, interspecific hybridization, and whole genome doubling.

The one issue that has effectively been settled in a convincing way is the evidence for a process of evolutionary change over the past three billion years. The reason the answer to this question is so solid is that every new technological development in biological investigation—from the earliest days of paleontology through light microscopy and cytogenetics up to our current molecular sequence methodologies—has told the same story: living organisms, past and present, are related to each other, share evolutionary inventions, and have changed dramatically over the history of the Earth. However, little evidence fits unequivocally with the theory that evolution occurs through the gradual accumulation of "numerous, successive, slight modifications" [1]. On the contrary, clear evidence exists for abrupt events of specific kinds at all levels of genome organization. These sudden changes range from horizontal transfers and the movement of transposable elements through chromosome rearrangements to whole genome duplications and cell fusions. In this part of the book, we will search for alternative conceptual foundations that better account for our current knowledge of genome change over evolutionary time.

A Systems Approach to Generating Functional Novelties

The question at the beginning of this book was about how functional new adaptations arise in evolution. Partial answers come from the knowledge of molecular networks and genome organization accumulated over the past half century. In that period, as outlined in Part I, we have witnessed a paradigm shift in scientific thinking from an atomistic, mechanical, reductionist viewpoint to a systems perspective that incorporates cell circuitry and molecular networks into a more integrated view of cellular and organismal activities, based in large measure on information processing. Current "systems" thinking attributes primary functional significance to the collective properties of molecular networks rather than to the individual properties of component molecules [46].

How does our knowledge of heritable variation fit into this new perspective? The most basic answer is that hereditary change results from processes carried out by dedicated cell functions subject to regulatory architectures. Our view of genome change has become one that describes active cell processes rather than a series of random accidents. Today we think in molecular and cellular terms about change processes at all levels:

- Localized point mutations result from untemplated incorporations by mutagenic DNA polymerases.
- Organisms acquire new biochemical routines through horizontal DNA transfers.
- Major genome rearrangements result from repair of DS breaks or from the action of mobile genetic elements.
- Whole genome mergers are the products of phagocytosis or cell invasions leading to endosymbiosis.

The various natural genetic engineering operators described in Part II are the molecular agents of active genome change. What characteristics do these operators display that fit them into the new systems perspective? The brief answer is that they have the capacities needed for a process similar to the kind of engineering humans undertake when we want to develop novel products or carry out established functions in a more efficient or responsive manner. We do not continually reinvent the wheel or randomly tweak individual systems, as the traditional evolutionary theory postulates. Instead, we reassemble known system or circuit elements in novel ways and, when technological advance or human imagination makes them available, we introduce novel components to a system. The long route from the transistor and integrated circuits to smart phones is a relevant example. Accordingly, let us adopt systems engineering as a theoretical metaphor for evolutionary change. To justify this metaphor, we will list some empirically well-documented biological analogies to systems engineering capacities. Then we will move on to consider more speculative ideas about what additional capacities might be uncovered by further research.

Reorganizing Established Functions to Generate Novelty

When an electronic or microchip engineer sets out to design a new circuit or chip architecture, she bases it on existing components that can be assembled by available technology. This is analogous to utilizing existing evolutionary inventions as the basis for a novel genomic architecture: coding sequences (exons), regulatory signals, chromatin formatting signals, and higher-order constructs such as centromeres, telomeres, and Hox complexes. In Parts II and III, you saw how cells possess the molecular apparatus to amplify the corresponding sequence components as either DNA or RNA and then place them in new locations and new arrangements in the genome.

As described earlier for protein evolution by domain swapping, this combinatorial process has a far greater probability of success than trying to modify each element in the genome one nucleotide at a time. The genomic assembly mechanisms have sufficient flexibility that they can operate hierarchically to rearrange all types of successful evolutionary inventions. These DNA inventions can range in size and intricacy from a single exon encoding a protein domain to a complex enhancer element executing a sophisticated regulatory algorithm to a complete genetic locus encoding one or several alternative RNA and protein products and even to megabase-sized conserved (syntenic) chromosome segments.

The way that biological circuits operate is a powerful driver of evolution by amplification and reorganization of genome segments. Reliability is of primary importance in cell function, and complex redundant networks are far less sensitive to breakdown from damage to a single component than are highly efficient minimal machines [42, 1101]. To put the importance of reliability in subjective human terms, it is far more reassuring to fly in an airplane that we know has multiple backups in case of individual system failure than in one that lacks this redundancy.

There is an additional and fundamental biophysical reason for generating redundancy to be a key aspect of the evolutionary process. In Part I, you saw a few examples of how important cooperative synergistic interactions between repeated signals have proven to be. For example, *lac* operon control involves interactions between two palindromic (internally redundant) *lacO* operators and four copies of the LacI repressor protein chain [66, 67] (see cartoon in Appendix I.1 online). Even higher levels of operator-repressor repetition and cooperativity have been discovered in the equally paradigmatic lambda phage repression system [1102]. The need for cooperativity arises because many biomolecular interactions are either weak or transitory, and multiple synergistic events stabilize the formation of functional complexes for carrying out cellular tasks such as replication, transcription, and cell division. The benefits of numerous cooperative interactions mean that we should expect to find iteration of interacting molecular determinants, such as DNA-binding domains and protein recognition sequences. Such multiplicity is exactly what genomic analysis has documented, both in protein structure and in the arrangement of signals at critical regions of the genome.

Generation of Novel Components

Clearly, successful engineering of new functionalities depends on more than devising novel combinations of existing components. All sorts of factors drive the invention of new methods and new ingredients: availability of materials, advances in technology, contacts with outside cultures, and human ingenuity. Thus, if we are to pursue the systems engineering metaphor, we have to consider how new cellular and genomic components can arise. At the DNA level, the two alternative proposals for how this occurs are the traditional assumption—multiple small random (therefore, independent) changes at dispersed locations in the genome—and the natural genetic engineering processes described in Part II.

Looking at the genomic record in Part III, we saw how complex DNA structures evolved in well-defined molecular processes through the action of mobile genetic elements and the posttranscriptional RNA processing apparatus. Documented examples include the generation of novel exons, introns, cis-acting control regions, and control molecules such as ncRNAs. If we are looking for an evolutionary reason to explain the subdivision of protein-coding sequences into exons and introns, one good candidate would be the increased facility this kind of split organization (and the accompanying RNA processing apparatus) provides for the rapid generation of genomic innovations. In other words, as evolution proceeds, so does evolvability.

In a much-cited article, François Jacob, one of the authors of the operon concept, employed the term *tinkering* to describe the use of genomic fragments to generate new functional DNA sequences [1103]. While the term is appropriate in that it describes the adaptation of existing materials to new functions by experimentation, it does not indicate the systematic and pervasive way this process of sequence adaptation has occurred in the course of evolution. In addition, the idea of tinkering does not explain why the most abundant DNA in the largest genomes is precisely the raw material for the creative rearrangements that generate and then amplify novel sequence elements. That is why the term *engineering* seems to be more appropriate for the built-in processes of self-modification that have operated over the course of evolution.

In human engineering, innovation is greatly facilitated by the presence of a vigorous research and development (R&D) sector, exemplified by institutions such as Bell Labs, NIH, and the great research universities. The freedom to experiment without concern for immediate utility has proven to be extremely valuable in generating novelties. Is there any biological analogy for such an unfettered R&D process? A number of molecular biologists have pointed out that viruses are biological entities that reproduce in an opportunistic manner that is not subject to the same functional constraints as cellular proliferation. They have proposed that the virosphere represents the realm where experimentation with genomic processes is least restricted and that viruses serve as sources of novelties that can later be adapted by cells [1104–1112].

Retention, Duplication, and Diversification of Evolutionary Inventions

There is a clear trend in the fossil record toward increasing size and structural complexity of organisms over time. The earliest cells have left no discernable trace. Early fossilized communities of prokaryotes, such as stromatolites, can sometimes be sizable [1113] [1114], but we know that the appearance of eukaryotes marked a quantum leap in cell and, ultimately, organism complexity. The early multicellular fossils of marine animals from the Ediacaran are smaller and have simpler organizations than their Cambrian successors, and the successful colonization of land by plants and animals has led to the appearance of even larger and more intricately organized forms.

Rather than attribute this apparent pattern of increase in size and complexity to some undefined innate drive, as Lamarck did (www.ucl. ac.uk/taxome/jim/Mim/lamarck_contents.html), we can understand the succession of larger and more intricately organized forms over time to illustrate the tendency of living organisms to retain, amplify, diversify, and reuse their evolutionary inventions. This growth in complexity occurs not so differently from the way that human engineers continually find new ways to build upon and add to technological innovations. We can list (in increasing order of complexity) just some of the evolutionary inventions discussed in Parts I to III that have been subject to widespread reuse and readaptation:

- Among proteins, DNA-binding domains, protein domains assembled into transcription factors, G factor-coupled receptors, and MAPK kinase signaling cascades.

- Among DNA elements, oligonucleotide recognition sequences, these sequences assembled into enhancers, centromeres, telomeres, and Hox complexes.

Any knowledgeable cell biologist or Evo-Devo specialist could easily expand the list of individual protein and DNA inventions. The end results of all this evolutionary creativity include such wonders as flowering plants and bilaterally symmetrical animal body plans.

The significance of this clear pattern of retention, amplification, and readaptation is that the organisms presently on Earth—namely, the organisms that have succeeded over evolutionary time—possess the natural engineering systems needed to duplicate and modify increasingly complex genomic constructs. It requires great faith to believe that a process of random, accidental genome change could serve this function. Indeed, as many biologists have argued since the 19th Century, random changes would overwhelmingly tend to degrade intricately organized systems rather than adapt them to new functions. Thinking about how many steps are needed to amplify and modify any complex genomic subsystem, the advantages of the division of restructuring capacities become increasingly apparent: DNA elements can rearrange large chromosome segments, and retroelements can retrotransduce smaller segments. The duplication and relocalization of a complex genomic structure extending hundreds of kilobase pairs fits the known capabilities of transposons, while the introduction of exons encoding novel interaction domains fits the processes experimentally demonstrated for non-LTR retrotransposons.

The Implications of Targeting Genome Restructuring

There is a convincing (perhaps overwhelming) case for the role of basic engineering principles in genome evolution. We now have many clear examples of genome restructuring by natural genetic engineering functions. Nonetheless, the phrase *natural genetic engineering* has proven troublesome to many scientists because they believe it supports the Intelligent Design argument. As one Nobel Laureate put it after a seminar, "If there is natural genetic engineering, that means there has to be an engineer." This empirically derived concept seems to many scientists to violate the principles of naturalism that exclude any role for a guiding intelligence outside of nature. Let us examine these critiques by asking the following questions:

- Can a distinct evolutionary advantage be gained by targeting genome restructuring?
- Can we envisage reasonable cellular and molecular mechanisms for heuristic genome targeting to occur?
- Is the intelligent (or "thoughtful," to use McClintock's term [644]) application of such molecular mechanisms outside the boundaries of contemporary biology?

Can a Distinct Evolutionary Advantage Be Gained by Targeting Genome Restructuring?

As we have seen throughout our discussions of cell metabolic regulation, cell reproduction, intercellular signaling, and multicellular development, the current systems biology view is that functionality resides in the correct operation of molecular networks more than it does in the action of individual molecules. This means that the synthesis and operation of molecules within a functional network are most efficient when they are coordinated. Biologists have repeatedly documented molecular features that enhance coordination (such as shared transcriptional controls, shared protein-protein interaction domains, and shared regulatory circuits).

From a systems perspective, it is obvious that natural genetic engineering processes will have the highest probability of generating successful novelties if they can recognize regions of the genome encoding functionally related molecules and produce similar or complementary changes in those regions concurrently. For example, changing the inputs or outputs of a cell network may involve inserting the same protein domains into several different molecules that work together to receive or transmit signals or placing these molecules under the control of the same transcriptional regulatory circuit.

If the natural genetic engineering functions that insert *cis*-regulatory signals and swap exons can operate simultaneously on more than one genetic locus encoding functionally related proteins, the chances of generating a workable invention will be greatly enhanced. The potential for such multilocus targeting will be especially great after a whole genome duplication event, when extra copies of the multiple loci encoding the components of many different networks are available for rearrangement. If related functional components could undergo similar changes simultaneously, the evolutionary process would operate far more efficiently.

Can We Envisage Reasonable Cellular and Molecular Mechanisms for Heuristic Genome Targeting to Occur?

Part II documents numerous cases in which well-characterized nucleic acid and protein interactions target particular natural genetic engineering processes to different genomic locations (Table II.11). In the adaptive immune system, you saw how targeted genome change (somatic hypermutation, immunoglobulin class-switching) can be coupled to the transcriptional control apparatus (also Appendix II.2 online). The ciliated protozoa illustrate how cellular RNAs can serve to guide specific kinds of genome reassembly operations. P element homing demonstrates that distinct genome regions subject to common regulation can be preferential sites of insertion. So there is no shortage of molecular mechanisms that can be co-opted as heuristics for targeting natural genetic engineering operations.

In addition, we are learning how genetic loci undergoing specific functions (replication, transcription, repair) localize into special subnuclear compartments. Growing evidence suggests that functionally related loci are subject to active colocalization within the same compartment [312]. So there is also a plausible mechanistic basis for grouping such loci physically and temporally into a region where they can undergo similar natural genetic engineering processes, such as regulatory site insertion or exon swapping. Thus, we see there is nothing magical or implausible in thinking about how cells can be capable of introducing coordinated changes into different but functionally related regions of their genomes.

Is the Intelligent Application of Such Molecular Mechanisms Outside the Boundaries of Contemporary Biology?

Although they may go through many trial-and-error steps, human engineers do not work blindly. They are trying to accomplish defined functional goals. Can such function-oriented capacities be attributed to cells? Is this not the kind of teleological thinking that scientists have been taught to avoid at all costs? The answer to both questions is yes.

We began Part I with the statement that living cells do not act blindly; then we proceeded to describe examples of their sensory and regulatory capacities. The more we learn about the detailed molecular operation of cells, the more we appreciate the depth of the circuitry they contain to ensure the accurate, "well-informed" execution of complex functions. This is now the prevailing view in many fields of biology. The concept of checkpoint controls is ubiquitous in cell biology, the field of "plant neurobiology" is emerging [1115–1118], and the August 2010 issue of the journal *Nature Immunology* was dedicated to "decision-making in the immune system." Thus, it appears that the idea of cellular cognition and decision-making with well-defined functional objectives has gone mainstream [1119]. Even Darwin entertained similar ideas, comparing the searching action of root tips to the operation of an animal brain [1120, 1121].

From the foregoing, then, it should be evident that the concept of cell-guided natural genetic engineering fits well inside the boundaries of 21st Century biological science. Despite widespread philosophical prejudices, cells are now reasonably seen to operate teleologically: their goals are survival, growth, and reproduction. In multicellular organisms, cells have elaborate control regimes to ensure that they fit into the overall morphology and physiology. Antibody-producing B lymphocytes are only one of hundreds of dedicated cell types in a mammalian body that operate for specific functional ends. In the service of their goals, cells expend a great deal of their available energy and matter on information processing and regulation. All the regulatory RNA molecules we have come to recognize in the past 15 years are produced at the expense of large quantities of ATP and other nucleotide triphosphate high-energy compounds.

Besides fitting within the currently defined boundaries of biology, the concept of functionally targeted and coordinated natural genetic engineering is open to experimental test. The problem can be approached from two directions. The bottom-up approach is to design experimental systems in which cells have to produce two or more targeted changes to pass through a particular selection regime (such as growth on a medium that lacks a particular nutrient or that uses a novel growth substrate). The frequencies of single and multiple change events can be determined, and the roles of particular natural genetic engineering systems, such as transposons and retrotransposons, can be evaluated. In this way, we can develop an appreciation of the potential for coordinated genomic changes. The successful isolation of bacterial mutants containing multiple related mutations within periods as short as a month is a promising indicator that the experiments can prove successful [1122, 1123].

The top-down approach to investigating targeted changes in the genome is to utilize our rapidly growing ability to obtain and interpret whole genome sequences [1124]. We can subject various organisms to particular kinds of "genome shocks" known to activate complex genome restructuring (Tables II.7 and II.8) and then carry out complete genome sequencing of the survivors that display novel heritable characteristics. The results will indicate the nature of the genome changes that have produced the new traits, and we can see how many of them have involved coordinated changes in network function. This method should be particularly applicable in well-studied plant species, such as *Arabidopsis thaliana*, where evidence already exists for morphological, chromosomal, and transcriptional changes following interspecific hybridization [1083, 1125–1128].

If the ideas of cell cognition, decision-making, and goal-oriented function are within contemporary biological perspectives—and if the natural genetic engineering concept is subject to empirical investigation—we can legitimately ask why the idea has been so fiercely resisted by mainstream biologists, and evolutionists in particular. My personal opinion is that the opposition is deeply philosophical in nature and dates back to late 19th Century disputes over evolution and also to the early 20th Century "mechanism-vitalism" debate [1129–1131]. The notion that random, undirected processes fully characterize natural systems (as they do in theoretical thermodynamics) was uncritically accepted at those times in much of the biological community. Over time, it came to be unchallenged conventional wisdom that cognitive, goal-oriented processes have to be relegated to the realms of unscientific fancy and religion.

The random process perspective expanded with the mid-20th Century shift toward molecular biology and away from examining integrated organismal functions. Molecular biologists often have been asked if they have discovered any fundamental new scientific principles at work in living organisms [1132], and the answer is a virtually unanimous no. Basic physics and chemistry allow us to dissect cells and identify molecules—even to determine their structures and explain many of their activities mechanistically and dynamically. Thus, according to this majority view, nothing else is needed. As mentioned in the Introduction and Part I, this fixation on random processes started to give way when molecular methods began to illuminate the operation of cell control regimes—and as molecular biologists increasingly turned their attention to complex phenomena such as immunity and multicellular development. Meanwhile, cybernetics [1133], control theory (http://www.cds.caltech.edu/~murray/papers/2003a_mur+03-csm.html), computation [1134], and electronics were developing entirely new and rigorously scientific ways to think about cognition, decision-making, and goal-oriented function. Thus, at the start of the 21st Century, we have a radically different conceptual environment, and the time has arrived to rescue evolution science and reintegrate it into these contemporary intellectual trends.

Can Genomic Changes Be Linked to Ecological Disruptions?

Among the most striking features of the fossil record are the periods of accelerated mass extinctions followed by periods of accelerated mass "originations" (appearances of morphologically novel organisms). The most famous of these episodes is the event that led to the disappearance of dinosaurs at the end of the Cretaceous period 65 MYA. As Luis Alvarez, his son Walter, and colleagues predicted in the 1980s, this event has fairly definitively been attributed to a large meteorite striking the Yucatan peninsula and distributing particulate matter worldwide to create a global ecological crisis **[1135]** [1136, 1137].

In the 1980s, Jack Sepkowski and David Raup systematically collected quantitative data on extinctions of genera in the Phanerozoic era, when animals were large enough to leave readily distinguishable fossils [1138–1140]. These mass extinctions occurred at the boundaries between one geological age with its characteristic fauna and another with a different fauna. Raup and Sepkowski first documented the "big five" mass extinctions in 1982:

Geological Boundary	Date or Interval (MYA)
Ordovician-Silurian	~450
Devonian-Carboniferous	~370
Permian-Triassic	251
Triassic-Jurassic	205
Cretacious-Tertiary	65

Since then, mass extinctions have been identified at other major geological boundaries [1141, 1142]:

Geological Boundary	Date or Interval (MYA)
Ediacaran-Cambrian	~542
Cambrian-Ordovician	~488
Silurian-Devonian	~416
Jurassic-Cretaceous	145.5

Originations of new groups and rapid diversification (or *radiation*) of certain groups, such as *Bilateria* in the Cambrian, characteristically follow mass extinctions **[1045, 1046]**. These rapidly radiating groups may have arisen in the previous geological period, as we now know was the case for *Bilateria* [1044]. There is general agreement that depopulation of existing ecological niches and the creation of new ecological niches by the novel organisms themselves are critical but hard-to-quantify factors in the dynamics of the appearance of new organisms in the record. Over time, of course, originations have to outnumber extinctions for life to sustain and increase its necessary diversity. There is a long-term increase in the number of taxonomic families visible in the fossil record that extends over multiple mass extinction-origination episodes [1143–1145].

Darwin thought that unfilled gaps in the fossil record could explain why it did not show his predicted pattern of gradual transformation of one life-form into another. His uniformitarian philosophy did not allow him to consider the role of extraordinary catastrophic events leading to mass extinctions. We do know that gaps exist in the fossil record because we have examples of three successive strata where fossils of a particular life-form are present in the oldest and most recent strata but cannot be found in the middle one [1146, 1147]. It turns out, however, that gaps tend to make the fossil record appear smoother than it actually is. A gap makes an extinction event seem earlier than it occurred and makes an origination or radiation event seem later. Statistical correction for gaps, therefore, actually accentuates the episodic and discontinuous nature of the fossil record (M. Foote, unpublished presentation at the University of Chicago, March 3, 2010).

How are we to integrate the known facts about the processes of genome change into a long-term perspective considering the changing history of the planet and the biosphere, and allowing for unpredictable events such as the meteorite impact at the end of the Cretaceous? Much attention has been devoted to analyzing the effects of atmosphere, climate, and temperature changes on the evolutionary record [1148, 1149]. One of the major features of biological origin involves changes in the gas composition of the atmosphere. These changes are largely a consequence of microbial evolution affecting the processes of methanogenesis, nitrogen fixation, nitrification (oxidation of ammonia to nitrate), denitrification (conversion of oxidized forms of nitrogen to N_2 gas), two-stage photosynthesis liberating O_2, and aerobic metabolism to generate CO_2. The September 1970 issue of *Scientific American* is devoted to biogeochemistry; a number of articles from that issue are listed in the Suggested Readings online. The microbe-generated atmospheric transformations have had major effects on the larger organisms, which left fossils, by setting the boundary conditions for physiological activity.

Although high-level changes in the biosphere have been considered [1150], little attention has been paid to the relationship between ecological disruption and genetic change. The influence that stimulus-sensitive regulatory processes and changes in population structure may have on the processes of genome restructuring requires greater scrutiny. If we think about the kind of ecological disruption that will lead to mass extinctions, it is readily apparent that many organisms will suffer severe physiological and nutritional distress and that populations will be greatly depleted, thereby altering customary mating patterns and host-symbiont relationships (commensal, pathogenic, parasitic, and mutualistic). The population depletion effect can reasonably be expected to lead to a marked increase in abnormal matings between organisms previously subject to reproductive isolation mechanisms and also to unusual associations between hosts and novel infective agents (symbionts and pathogens). As we saw in Part II, these are precisely the conditions that stimulate natural genetic engineering functions (Tables II.7 and II.8). Hybrid dysgenesis, interspecific mating, infection, and prolonged stress conditions in the laboratory can all serve as experimental proxies for the genome-destabilizing events that would logically follow major ecological disruption.

What Might a 21st Century Theory of Evolution Look Like?

Conventional views of evolution were formulated before we learned about the structure of DNA and embarked on the molecular analysis of cells, morphogenesis, and the genomic record. They were formulated in the mid-19th Century and then reformulated in the mid-20th Century, when the prevailing philosophy in science was characterized by atomistic, mechanistic, and statistical thinking. The basic elements of the Modern Evolutionary Synthesis included an *ad hoc* assumption about the random nature of hereditary variation, the diversifying effects of Mendelian segregations according to the rules of quantitative population genetics, the positive action of natural selection, and a belief that geological time was sufficient for the selection-guided accumulation of small adaptive changes to produce new life-forms [190]. As mentioned earlier, the early pioneering accomplishments of molecular biology were interpreted to provide a solid physical and chemical basis for this perspective.

Today, we have a different body of information and a distinct scientific mind-set. We know about cellular information processing and control circuits, the ability of cells to repair and restructure their genomes, and the rapid large-scale changes in genome organization that distinguish life-forms. We know about three different domains of cellular life, the evolutionary importance of horizontal DNA transfers, and multiple cell mergers leading to endosymbiosis and phylogenetic diversification. We see the fossil record as episodic and characterized by geologically abrupt changes in the nature and distribution of organisms. Contemporary science has been deeply influenced by developments in electronics, computation, and the informatic sciences. The tools these areas have provided allow us to focus on complex interactive systems with decidedly nonlinear characteristics. Our most challenging problems, such as climate change, oblige us to integrate multiple distinct feedbacks as central to our analytical methods.

Using this 21st Century scientific perspective, we can articulate a more interactive and information-based set of basic evolutionary principles without departing from the realm of established empirical observations:

- Living cells and organisms are cognitive (sentient) entities that act and interact purposefully to ensure survival, growth, and proliferation. They possess corresponding sensory, communication, information-processing, and decision-making capabilities.

- Cells are built to evolve; they have the ability to alter their hereditary characteristics rapidly through well-described natural genetic engineering and epigenetic processes as well as by cell mergers.

- Evolutionary novelty arises from the production of new cell and multicellular structures as a result of cellular self-modification functions and cell fusions. In many cases, these new structures involve the amplification and/or rearrangement of existing functional molecular components. In addition, cells (and associated entities such as viruses) have the capacity to generate entirely new nucleic acid, protein, and other molecular components that can subsequently be integrated into functional cell or multicellular systems. As discussed previously, it remains to be thoroughly investigated to what degree cell sensing, information processing, and genome targeting can heuristically accelerate the production of useful novelties. *A priori*, functionally relevant guidance for natural genetic engineering would appear to enhance the probability of success in generating useful novelties. This supposition requires rigorous testing.

- Natural genetic engineering and other evolutionarily innovative processes respond to stimuli that place the core organism objectives of survival, growth, and proliferation in peril. These dangerous challenges are most likely to occur at times of major ecological disruption.[1]

- Active hereditary variation and evolutionary innovation will continue as long as ecological disruptions and challenges to core vital objectives persist.

- The role of selection is to eliminate evolutionary novelties that prove to be non-functional and interfere with adaptive needs. Selection operates as a purifying but not creative force.

- Evolutionary inventions that survive purifying selection and prove useful are subject to microevolutionary refinement, perhaps by the kind of processes envisaged in conventional theories.

- Successful evolutionary inventions are subject to amplification, reuse, and adaptation to new functions in response to successive ecological challenges.

- Taxonomically specific characters become established as the functional integration of past evolutionary novelties increasingly places constraints on the kinds of additional inventions that will prove functional.

An evolutionary process that follows these principles will naturally display the kind of episodic and abrupt changes found in both the fossil and genomic records. Indeed, the *punctuated equilibrium* pattern of Gould and Eldridge should be the default situation, because the stimulating role of ecological challenges is unpredictable **[1151]** [1152–1154]. When contemporary knowledge of natural

1 In her 1983 Nobel Prize address, Barbara McClintock made the following prediction: "In the future, attention undoubtedly will be centered on the genome, with greater appreciation of its significance as a highly sensitive organ of the cell that monitors genomic activities and corrects common errors, senses unusual and unexpected events, and responds to them, often by restructuring the genome" [644].

genetic engineering and horizontal transfer is incorporated into the analysis, for example, it has recently been possible to calculate from genome databases a striking episode of rapid evolutionary innovation in the Archaean eon (~3,200 MYA) [1155].

Where Does Evolution Fit in 21st Century Science?

The science of the 21st Century deals with the interactions between the multiple components of complex systems, ranging from aggregates of elementary particles (each of which has its own multivalent set of properties) to the behavior of the largest structures in the cosmos. This kind of science is fundamentally different from earlier periods, when the goal was to understand the unique property of each atomistic unit and then try to derive the behavior of large systems from a small set of interaction rules plus the character of their component parts. Today, a major focus in scientific inquiry is to understand how systems change over time, whether they are atoms, molecules, organisms, ecosystems, climates, galaxies, black holes, or universes.

Change over time—in other words, evolution—is thus a central feature of contemporary science. Biological evolution has long served not only as a metaphor but also as a model for the other sciences. The dominant atomistic ("gene"-based) and statistical (mathematical population genetics) approaches to biological evolution have been enormously influential in many fields. Indeed, the founders of mathematical population genetics (R. A. Fisher, J. B. S. Haldane, and S. Wright) are considered major figures in the development of statistics and in its application to all fields of physical and social as well as biological science. In addition, biological evolution has served as a model for developments in the information sciences, such as the elaboration of "Genetic Algorithm" methods by John Holland and his followers [1156] (http://www2.econ.iastate.edu/tesfatsi/holland.gaintro. htm) as well as the fledgling field of evolutionary computation [1157–1159].

Given the exemplary status of biological evolution, we can anticipate that a paradigm shift in our understanding of that subject will have repercussions far outside the life sciences. A shift from thinking about gradual selection of localized random changes to sudden genome restructuring by sensory network-influenced cell systems is a major conceptual change. It replaces the "invisible hands" of geological time and natural selection with cognitive networks and cellular functions for self-modification. The emphasis is systemic rather than atomistic and information-based rather than stochastic.

How such an evolutionary paradigm shift will play out in the physical and social sciences remains to be seen. But it is possible to predict that the cognitive (psychological) and social sciences will have an increased influence on biology, especially when it comes to the acquisition and processing of information. Parallels have long been noted between linguistics and genome expression, but the evolutionary perspective advocated here will make it much easier to incorporate lessons about language evolution into theories of organic evolution [1160, 1161]. The conceptual change should also temper theorizing in psychology and linguistics based on 20th Century genetic determinism [1162]. It is not inconceivable that parallels will be established even between disciplines as seemingly distant as cell biology and economics. For example, the role of cell cycle controls and checkpoints can be taken as models for market rules and government regulation, and the control breakdowns characteristic of cancer cells may serve as models for the information transfer defects that lead to market failures, such as the recent crash of 2008.

The interest of physical scientists such as Max Delbrück and Jean Weigle in the 20th Century had a revolutionizing effect on biology. Focusing on the molecular basis of vital functions transformed the life sciences and, ultimately, made us aware of the intricate circuitry that governs cell reproduction, multicellular development, and genome reorganization. Currently another wave of physical scientists is entering the life sciences. They bring with them a much-needed and fruitful sophistication in observation at the micro level, in mathematical formalizations of results, and in computational methods of data analysis. Physicists-turned-biologists have the additional advantage of lacking a formal education in the life sciences; consequently, they have not been taught to exclude from their thinking notions previously concluded to be "impossible." We can only hope that their less prejudiced backgrounds will make it easier for them to develop novel conceptual frameworks to complement the analytical and experimental techniques they are introducing.

Evolution is life's way of dealing with the unpredictable. We have seen that principle most clearly at work in the adaptive immune system, where antibodies have to be synthesized that can recognize unknown invaders. The fact that future adaptive needs are unknowable does not mean that filling those needs has to be a blind process. In immune system natural genetic engineering, and in evolutionary change in general, we have been able to discern regular features of genome restructuring that facilitate the production of novel molecular tools with an enhanced likelihood of real-world utility. A measure of success for the more informational perspective sketched out in this book will be the extent of future research into the cognitive cellular operations that have led to successful evolutionary inventions. We have a great deal to learn in this respect.

Evolving life has far exceeded human ingenuity in producing immensely complex and reliable self-reproducing entities that have repeatedly managed to change, survive, and proliferate despite major ecological upheavals. Given the challenges we face as a species, it behooves us to find out as much as we can of nature's wisdom in dealing with the inescapable trials of life.

Glossary

7S RNA
A stable cell RNA transcribed by RNA polymerase III, a component of a particle that guides newly synthesized proteins across membranes.

Ac-CoA
Acetyl coenzyme A, a small molecule that is used to transfer acetyl groups to proteins and other molecular targets.

acetyl group
-$COCH_3$, a two-carbon chemical group used to modify proteins and other molecular targets.

adaptive immunity
The ability to recognize and maintain the capability to inactivate or destroy specific infectious agents.

allotetraploid
An organism with two diploid genomes that come from different parents.

angiosperm
A flowering plant that encloses its seed in a fruit.

anneal
In molecular biology, the base-pairing of two complementary nucleic acid strands to form a duplex structure.

annotation
Interpretation of nucleotide sequence data to indicate where its coding, signaling, or other functional capacities are located.

antibody
An immune system protein produced by B cells that binds to specific structures on infectious agents and other foreign bodies in vertebrates. Composed of two longer (heavy) and two shorter (light) chains.

antigen
A structure that is recognized and bound by an antibody.

antigenic variation
The ability of an infecting cell to change the structure of surface molecules recognized by the immune system.

AP
Anterior-posterior (front to rear). Indicates the main axis of a bilaterally symmetric animal body.

apoptosis
Literally, "falling away." Programmed cell death.

apoptosis cascade
The complex sequence of molecular events that result in an organized process of programmed cell death.

ars
Autonomous replication sequence. The site where DNA replication begins in eukaryotic chromosomes.

autotetraploid
An organism that has two sets of diploid chromosomes of the same origin.

bacteriophage (phage)
Literally, "bacteria eater." A virus that infects and reproduces in bacteria.

Bilateria
Bilaterally symmetric animals.

bioinformatics
Computer analysis of protein and nucleic acid sequence data.

boundary element
A DNA signal that separates one genome domain from another (for example, separating regions of different chromatin configuration). Also prevents one transcription signal from interacting with another.

bp
Base pair. The minimal component of a DNA duplex.

breakpoint
A site of a chromosomal rearrangement, where two broken chromosome ends have joined.

cAMP
Cyclic adenosine monophosphate. A second messenger signaling molecule derived from adenosine triphosphate (ATP).

Cartesian dualist
Someone who believes in separating information processing from the execution of physical or chemical activities.

cassette
In molecular genetics, a segment of DNA containing a limited region of coding sequence, typically for all or part of a protein.

centromere
Literally, "central structure." The DNA region formatting eukaryotic chromosome interaction with the apparatus for proper movement at cell division.

centrosome/centriole
Literally, "central body." The structure in animal cells that nucleates and organizes the microtubule network that moves chromosomes during cell division.

chloramphenicol
An antibiotic that reversibly interferes with bacterial protein synthesis on the ribosome.

chromatin
Literally, "colored material." The combination of DNA, protein, and RNA that can be stained (colored) and visualized in the microscope.

chromodomain
A polypeptide region of 40 to 50 amino acids commonly found in proteins that bind to chromatin.

chromosome
An extended DNA structure carrying multiple genetic loci capable of replication and transmission to progeny cells during cell division. Frequently circular in bacteria.

chromosome painting
In molecular cytology, the use of fluorescent RNA or single-strand DNA probes to detect specific chromosomes or specific chromosome components.

cilium (plural, cilia)
An external cell structure capable of undergoing an undulating, wave-like motion.

***cis*-regulatory module (CRM)**
An organized cluster of DNA signals that control transcription.

commensal
Literally, "dining together." An organism that shares nutrition with another.

competence
In molecular genetics, a cell's ability to take up DNA from the external environment and integrate all or some sequences into its own genome.

constant region
In immunology, the antibody protein chain segment that is the same from one antigen-specific antibody to another.

coordinated transcription
The simultaneous or sequential copying of different DNA sequences into RNA.

CRISPR
Clustered regularly interspaced short palindromic repeats. A DNA structure in prokaryotic cells that incorporates short fragments of invading DNA molecules so that RNA can be synthesized that interferes with the expression of that invading DNA.

CRM
Cis-regulatory module. An organized cluster of DNA signals that control transcription.

CRP (*crp*)
cAMP receptor protein. A molecule that binds cAMP and the *crp* site in DNA to regulate transcription, generally to stimulate expression.

cut-and-paste transposition
The movement of a defined DNA segment (transposon) by double-strand breaks and insertion at a new site without replication.

cyanobacteria
Formerly "blue-green algae," a group of bacteria capable of oxygenic photosynthesis.

cytotype
The capacity of fertilized egg cells to control the activity of transposable elements introduced in sperm chromosomes.

diploid
A cell with two copies of its genome.

diploidization
The process whereby a polyploid genome loses chromosome segments to return to a diploid state.

dispersed repeat
A repeated DNA sequence element where the individual copies are found in different genomic locations.

DIVAC
Diversification activator. A DNA segment that inhibits error-free DNA repair and facilitates somatic hypermutation of antibody variable region coding sequence.

diversity-generating retroelement (DGR)
A bacterial system involving a reverse transcriptase protein and a repeated DNA cassette for changing the variable region of a nearby coding sequence.

domain
A segment of nucleic acid or protein with its own structure and functional properties.

DSB
Double-strand break in DNA.

duplex
A region of double-helical nucleic acid.

ectopic
Outside the usual location.

EGF
Epidermal growth factor. A signaling molecule in animals that stimulates cell proliferation.

endonuclease
An enzyme that makes an internal cleavage in a nucleic acid.

endoreplication
DNA replication in the absence of nuclear or cell division, amplifying the number of copies that frequently remain aligned.

endosymbiosis
The stable proliferation of one organism inside the cells of another.

enhancer
In molecular genetics, a sequence or collection of sequences that stimulates or otherwise modulates transcription.

epigenotype (epigenome)
The constellation of epigenetic (chromatin) configurations within the entire genome.

erythromycin
An antibiotic that blocks bacterial protein synthesis on the ribosome.

euchromatin
A form of chromatin that stains lightly and is considered to include actively transcribed or replicating DNA.

eukaryote
Literally, "true kernel." An organism whose cells have a defined nucleus, with a nuclear envelope surrounding its genome.

exchange
In molecular genetics, the transfer of DNA duplex regions between molecules.

exon
Literally, "expressed element." A DNA sequence segment, typically protein-coding, that ends up in the final RNA product following splicing.

exonuclease
An enzyme that cleaves nucleic acid from the end.

expression site
In molecular genetics, a genomic location where coding sequences are transcribed into functional messenger RNA molecules.

formatting
Borrowed from computer science, the addition of signals to genomic sites that indicate how they will interact with cellular molecules for functions such as chromatin formation, replication, transcription, and recombination.

gamete
A haploid sex cell. In higher eukaryotes, typically either a sperm or an egg.

genome
The complete collection of DNA molecules within a cell or other biological entity, such as a virus.

genotoxicity
The capacity to damage or chemically modify DNA.

genotype
In classical genetics, the complete collection of genetic determinants in a cell or organism.

germline
The lineage of cells ultimately destined to undergo meiosis and produce gametes.

germ plasm
An idea of August Weismann that the tissue comprising the germline is fundamentally different and isolated from other tissues.

gonococci
The coccus (spherical) form bacteria that cause gonorrhea.

gradualism
The concept that major change occurs slowly through the accumulation of small variations.

hairpin end
In immunogenetics, a structure formed at the site of a double-strand break when the two strands are joined at their extremities by a phosphodiester bond.

haploid
A cell or organism having a single copy of its genome.

helitron
A DNA transposon that uses a process of rolling circle replication in its movement from the donor to target site.

heterochromatin
Darkly staining ("heterochromatic") chromosome regions, generally considered to be inactive for transcription and replication but capable of playing special roles in chromosome placement or movement at cell division.

histone
One of a group of specialized protein molecules that bind and wrap the DNA into nucleosomes, the most basic step in DNA compaction in eukaryotic cells. Also the targets of chemical modifications in distinct forms of chromatin.

homeodomain
A DNA-binding domain found in many animal transcription factors.

homologous
Having or using homology.

homologue
One of a pair or a greater number of molecules that display homology. In genetics, applied to copies of the same chromosome.

homology
In biology, structural or sequence similarity, generally considered to indicate common ancestry.

homopolymer
A nucleic acid chain (polymer) composed of a single nucleotide repeated many times.

host organism
In infection and symbiosis, a large organism that harbors a smaller organism.

hox
Homeobox. The sequence encoding a homeodomain, applied to proteins and their encoding DNA sequences.

hox complex
An oriented cluster of homeodomain protein coding sequences active in executing AP differentiation in *Bilateria*.

HR
Homologous recombination.

hybrid
A cell or organism having two parents belonging to different species or populations.

hybrid dysgenesis
A syndrome of germline abnormalities and chromosomal changes observed among the progeny that result from matings between members of different populations.

hypersensitive response
In plants, the occurrence of programmed cell death surrounding an infected cell to create a barrier zone of dead tissue, preventing further spread of the infective agent.

hypha
A filament, typically applied to the extensive filamentous growth of fungi.

ICE
Integrated conjugative element. A large segment of DNA that encodes systems for its horizontal transfer between prokaryotic cells and insertion into the genome of the recipient cell.

IES
Internal eliminated sequence. In ciliate macronuclear development, a segment of the germline chromosomes that will be excised and discarded, absent from the final active genome.

IGF
Insulin-like growth factor. A signaling protein that stimulates mammalian cell growth.

imprinting
The epigenetic phenomenon whereby the chromatin and expression state of a chromosome region depends on the gender of the parent from which it is inherited.

insulator sequence (insulator body)
A signal for separating two distinct chromatin or transcriptional regions. In the cells of some organisms, multiple insulators can be observed to cluster in "bodies" visible at the nuclear periphery.

integrase
In molecular genetics, a protein that inserts specific DNA molecules into other genome molecules.

intein
A self-splicing internal segment of a protein. Frequently inteins acquire endonuclease activity when they excise from the parent protein.

interchromosomal exchange
Swapping of DNA segments between distinct chromosomes.

interphase
The periods of the eukaryotic cell cycle where the chromosomes are not condensed and are not clearly visible in the light microscope.

interspecific hybridization
Mating between parents of distinct species to produce a hybrid offspring. A frequent source of genome change and formation of new species.

introgression
The introduction of genetic material from one species into the genome of another species.

intron
Literally, "inserted element." A DNA segment whose RNA copy is removed from the final RNA molecule in the splicing process.

isotype
The character of an antibody molecule, determined by the constant region of its heavy chain components.

junctional diversity
In immunology, the ability of cells to join the same two DNA cassettes at any one of a number of different internucleotide positions, thereby creating diverse sequences from the same starting material.

kinetochore
The structure that forms upon the centromere sequence and physically connects the chromosomal DNA to the microtubules of the mitotic apparatus.

ligation
In molecular genetics, the joining of two nucleic acid strands, generally DNA.

LINE element
Long interspersed nucleotide element. A family of dispersed repeats that are also non-LTR retrotransposons encoding the proteins needed for movement to new genomic locations by themselves, by SINE elements, and by cellular RNAs. Capable of downstream 3' retrotransduction.

linkage
In genetics, the connection between two genomic sites that biases them to moving together in recombinational exchanges. Hence, an inverse measure of genetic distance.

lipid
A hydrophobic ("water-fearing") molecule derived from fatty acids or other hydrocarbons.

locus
Literally, "position." In genetics, the place where mutations causing related phenotypes tend to cluster on a genetic map. In molecular genetics, the constellation of neighboring DNA elements that contribute to the expression of one or more structurally similar molecules.

LSD
Large-scale duplication. Refers to genomes that contain many duplicated regions.

LTR
Long terminal repeat. The directly repeated structure that is found at each end of a retroviral provirus or a related retrotransposon in the genome.

lymphatic tissue
Tissue of the lymphatic system that gives rise to the cells and molecules of the adaptive immune system.

lymphocyte
An immune system cell.

lymphokine
An immune system signaling molecule.

lysogeny
Literally, "producing lysis." The inherited capacity of a bacterial culture to produce bacteriophage particles.

macromolecule
In biology, a large molecule composed of many similarly connected subunits, as in nucleic acids, proteins, polysaccharides, and membrane lipids.

MB
Mega (million) base pairs.

MDS
Macronuclear destined sequence. In ciliate macronuclear development, the segments of the micronuclear germline chromosomes that remain in the functional macronucleus after massive DNA cutting and splicing.

MeCP2
Methyl CpG binding protein 2. A chromatin and transcriptional regulatory protein in nerve cells whose absence causes Rett Syndrome, a neurological deficiency.

meiosis
A double cell division process in sexually reproducing eukaryotes whereby a single diploid cell produces four haploid progeny. The basis of making spores and gametes.

membrane
A large structure composed of lipids and proteins that serves as a barrier for the free diffusion of soluble molecules. Typically it separates the space between subcellular compartments and organelles.

metazoa
Multicellular animals.

methylation
The process of attaching a methyl group to a molecule.

methyl group
$-CH_3$, a basic component of organic molecules that can be attached to chemically mark many sites on biological molecules, such as proteins and DNA.

microsatellite
A short segment of tandem repeated DNA. Used as a marker for forensic DNA analysis.

microtubule
A type of cellular filament that can expand and contract. Microtubules are particularly important in chromosome movements at cell division through attachment to the kinetochores.

miRNA
Micro RNA. Small molecules that play a wide variety of regulatory roles by recognizing complementary sequences in RNA or DNA.

mitochondrion
Literally, "threaded granule." The subcellular organelle that carries out the energy-released oxidative breakdown of sugars and other nutrients in aerobic eukaryotic cells.

mitosis
The process of eukaryotic cell division that maintains the same number of chromosomes in each daughter cell.

mitotic spindle apparatus
The system of filaments that converge on either side of the dividing nucleus and pull the chromosomes into the two daughter cells during mitosis.

MMR
Methyl-directed mismatch repair. A post-replication proofreading process that removes improperly incorporated nucleotides in newly duplicated bacterial DNA.

morphogenesis
Literally, "production of form." The process of development leading to organized structures in animals, plants, and microbes.

multidrug resistance plasmid
A circular self-replicating DNA molecule encoding mechanisms for resistance to two or more antibiotics. Frequently transmissible from one bacterial cell to another.

mutagenic
Capable of inducing heritable changes in DNA sequence.

mutator polymerase
A specialized type of DNA polymerase for replicating past chemical damage (lesions) in the parental DNA strand, characterized by a high frequency of sequence novelties in the newly synthesized strand. Also called "translesion polymerase."

mutualism
A regular interaction between two distinct organisms that benefits both of them.

MYA
Million years ago.

mycorrhiza
Literally, "fungus roots." Symbiotic fungi that provide root function for plants.

natural genetic engineering
The collective set of biochemical capabilities that cells have to restructure their genomes by cleaving, splicing, and synthesizing DNA chains, much as we do in modern biotechnology.

NHEJ
Nonhomologous end-joining. The process of directly connecting broken ends of DNA duplexes without a homologous template for recombination.

noncoding RNA (ncRNA)
An RNA molecule that does not code for protein.

nuclear pore
A structure in the nuclear envelope that allows the import and export of macromolecules such as proteins and processed RNA.

nuclease
An enzyme that cleaves nucleic acids.

nucleosome
Literally, "nucleus body." A structure composed of eight histone protein chains with about 147 bp of duplex DNA wrapped around them. The most basic form of DNA compaction in eukaryotic cells.

nucleotide
The basic unit of a nucleic acid polymer, composed of a nitrogenous base attached to a ribose or deoxyribose sugar that is, in turn, attached to a phosphate group.

oligonucleotide
A short stretch of polymerized nucleotides, either RNA or DNA.

ORC
Origin replication complex. A nucleoprotein complex that forms at the start site of eukaryotic DNA replication.

ori
Origin of replication. The start site of DNA replication in prokaryotes.

origination
An appearance in the fossil record of a morphologically novel organism.

palindrome
A sequence that reads the same in both directions. In DNA, an inverted repeat.

paramutation
Induction of an epigenetic "mutation" or loss of expression following pairing with the epigenetically silenced form of certain genetic loci. Does not involve any sequence change.

par site
A site involved in the partitioning of replicated DNA molecules to each of the daughter cells in prokaryotic cell division.

penicillin
An antibiotic that has a beta-lactam ring in its structure and inhibits bacterial cell wall formation, inactivated by beta-lactamase enzymes.

peptide attachment
A chemical linkage of a (poly)peptide to an amino acid side chain in a protein, a common form of protein modification.

peptide bond
The chemical linkage between adjacent amino acids in a polypeptide chain.

peptide excision
Cleavage of a protein chain and removal of a polypeptide segment.

phase variation
In bacteriology and serology, the appearance or disappearance of a specific antigen.

phenotype
One or more observable properties of a cell or organism, sometimes used as a collective term for all an organism's traits.

phosphodiester bond
The chemical linkage between the sugars in two adjacent nucleotides of a nucleic acid polymer.

phylogeny
Literally, "birth of a race." The pattern of descent of a group of related organisms or molecules.

piRNA
Animal germline small RNA molecules that bind to Piwi proteins and direct epigenetic silencing of complementary DNA regions.

Piwi protein
An RNA-binding protein that participates in piRNA-directed epigenetic silencing in the animal germline.

plasmagene
A hypothetical hereditary determinant in the cytoplasm of ciliate protozoa and invertebrates, now identified as an endosymbiotic bacterium.

plasmid
An independently replicating DNA molecule in prokaryotic cells.

plastid
A cellular organelle that carries out photosynthesis in eukaryotic cells, sometimes applied to related organelles that have lost photosynthetic capacity. Often equivalent to chloroplast.

ploidy
The number of copies of the genome in a cell.

plus strand
The strand in a duplex DNA coding region of the genome whose sequence corresponds to the RNA transcript. Also the protein-coding strand of RNA.

polymerize
To join multiple subunits into a chain, such as nucleotides in a nucleic acid.

polypeptide
A sequence of amino acids connected by peptide bonds in a polymer chain.

polyploid
An organism with more than two copies of the genome.

polytene
Literally, "many bands." Refers to chromosomes that have multiplied without separating from each other. Found in special tissues such as insect salivary glands.

position effect
A change in the expression of a genetic locus resulting from a change in its genomic location.

post-transcriptional
Following transcription. Generally refers to the processing or modification of primary RNA copies of a genetic locus.

processed RNA
RNA molecules that have been modified by splicing, cleavage, or other chemical alteration.

programmed cell death
A regular process of cellular self-destruction, characterized by well-defined biochemical steps and subject to predictable control stimuli.

prokaryote
Literally, "before kernel." An organism that lacks a defined nucleus surrounded by an envelope.

promoter
A DNA signal that marks the site where transcription begins.

prophage
The latent but inherited form of a bacteriophage genome, either integrated into the host prokaryote genome or replicating as a plasmid.

protein
A macromolecule composed of amino acids polymerized into one or more polypeptide chains.

proteolytic cleavage
A break of the peptide bonds in a protein chain.

protist
A eukaryotic microorganism.

RAD9
An *S. cervisiaea* checkpoint protein that participates in halting the cell cycle prior to division when there is DNA damage.

RAG proteins
Recombination activation gene proteins RAG1 and RAG2. Responsible for DNA breakage necessary for the joining of V, D, and J cassettes encoding segments of antibody proteins. Related to a family of transposases and able to carry out transposition reactions in the test tube and yeast.

RecA
A protein required for homology-dependent recombination, which also acts as a DNA damage sensor to active the bacterial SOS DNA damage-response system.

recombinase
A protein that carries out genetic exchange (recombination) between two DNA segments. Mostly applied to proteins responsible for site-specific recombination.

repetitive DNA
DNA sequences that appear many times in the genome.

replication
The process of duplicating double-stranded DNA into two new duplexes.

replication fork
The structure that forms at the site where the two strands of a replicating DNA duplex are separated for polymerization of new strands complementary to the two parental templates.

replicative transposition
A process of mobilizing DNA transposons to new locations in which transposon replication occurs.

replicon
An inherited replicating DNA molecule in a genome, generally applied to prokaryotes and nonchromosomal genetic elements in eukaryotes.

retroelement
A genomic DNA component produced by reverse transcription of an RNA molecule.

retrohoming
The process of inserting an intron into the genome by reverse transcription.

retrosplicing
The process of inserting an intron into the genome by a direct chemical reversal of the splicing reaction, with plus strand DNA taking the place of spliced RNA.

retrotransduction (retroduction)
The process of moving a DNA sequence from one location to another by means of an RNA intermediate. Typically applied to sequences adjacent to a retrotransposon.

retrotransposition
The movement of a genetic element to a new location in the genome by means of an RNA intermediate and reverse transcription.

retrotransposon
A genomic element that duplicates and moves to new locations by retrotransposition.

retrovirus
An RNA virus that reproduces by reverse-transcription, inserting its cDNA copy into a cellular genome, and then making new genomes by transcribing the inserted provirus.

reverse transcriptase
An enzyme that polymerizes DNA chains complementary to a parental RNA template.

ribosome
A cellular organelle composed of RNA and proteins in two subunits (large and small), where the messenger RNA sequence is translated into the amino acid sequence of a polypeptide chain.

ribozyme
A catalytic RNA molecule.

RNA loop
A structure that forms when RNA anneals with one strand of a DNA duplex and displaces the corresponding DNA strand.

RNA maturase
A protein that binds to and facilitates post-transcriptional processing of RNA molecules.

RNA PolIII
RNA polymerase III. The enzyme that transcribes small stable RNA molecules in eukaryotic cells.

rRNA
Ribosomal RNA. Stable RNA molecules (different for each ribosome subunit).

Saccharomyces cerevisiae
Budding or brewer's yeast, one of the principal model organisms for eukaryotic genetics.

saltationist
Occurring in leaps.

SAM
The small molecule S-adenosyl-methionine. Used to transfer methyl groups to proteins and DNA.

satellite DNA
DNA that has a different density than the bulk of the genomic DNA (it forms a *satellite* peak in density measurements). Now used to describe DNA having a tandem repetitive structure and different nucleotide composition than the rest of the genome.

Schizosaccharomyces pombe
Fission yeast, one of the principal model organisms for eukaryotic cell cycle studies.

second messenger
A small molecule that diffuses from one macromolecule to other macromolecules to carry information in cell signal transduction.

self-splicing
Catalytic RNA molecules that carry out their own splicing reactions.

signal transduction
The process of transmitting information within cells.

silencer
A genetic element that inhibits transcription.

simple sequence repeat (SSR)
Repetitive DNA that consists of many tandem copies of a short sequence, typically 2 to 6 bp in length.

SINE element
Short interspersed nucleotide element. A class of abundant retro-transposons in eukaryotic cells that do not encode the proteins for their own mobility.

siRNA
Small interfering RNA. Generally, a small sequence of about two to three dozen nucleotides that directs the formation of silent chromatin or interferes with messenger RNA function.

site-specific recombination
Reciprocal genetic exchange that occurs at two specific signals recognized by a cognate recombinase protein.

small molecule
A molecule that is not a polymer composed of many subunits.

somatic
Literally, "of the body"—that is, distinct from the germline. Refers to cells or tissues that reproduce by mitosis.

SOS response
A complex bacterial response to DNA damage that involves expression of repair, recombination, and cell division inhibition proteins.

spindle poles
The locations on opposite sides of the nucleus where microtubules originate to contact the pairs of duplicated chromosomes in the M phase of the eukaryotic cell cycle.

spirochetes
Bacteria with a spiral cell structure.

splicing
The process of removing intron sequences from an RNA transcript.

spore
The resistant, dormant form of various organisms, ranging from bacteria to ferns. In fungi, the haploid products of meiotic division.

SS
Single-strand.

Streptomycin
An antibiotic that irreversibly inhibits the ribosome in protein synthesis.

subnuclear
Comprising only a part of the nuclear volume.

SulA
A bacterial SOS protein that inhibits cell division until DNA repair is completed.

Sulfanilamide
An antibiotic that inhibits folic acid biosynthesis.

SVA element
A non-LTR SINE retrotransoposon composed of the following components: satellite sequence repeat $(CCCTCT)_n$ combined with VNTRs and Alu-like antisense segments. Capable of upstream 5' retrotransduction.

syncytial
Displaying a fused multicellular structure with many nuclei but only a single cell membrane.

syntenic
Literally, "on the same band." Colinear chromosome segments that share a common ancestral origin.

systems biology
A form of molecular biology that attributes functional significance to the collective properties of molecular networks rather than to the individual properties of component molecules.

tandem repeat array
Repetitive DNA sequences adjacent to each other, most commonly in a parallel head-to-tail arrangement.

target site
A genome location into which a mobile element inserts itself.

taxonomy
Classification. In evolution science, classifying molecules and organisms according to their inferred phylogenetic relationships.

T-DNA
Literally, "tumor DNA." The segment of DNA conjugated into a plant cell by *Agrobacterium* cells to direct the development of a cell mass that produces chemicals digestible by the Agrobacteria but not by other soil organisms.

telomerase
The enzyme that maintains and lengthens telomeres in most eukaryotic cells. A specialized kind of reverse transcriptase.

telomere
Literally, "end structure." The special DNA structure at the ends of chromosomes that permits elongation and blocks end-joining.

telophase
The subdivision of the M cell cycle phase when the duplicated chromosomes fully separate from their siblings and move toward the spindle poles.

temperate bacteriophage
A bacterial virus that can reproduce in one of two ways after infection. With lytic growth, viral reproduction kills the host cell and produces many new virus particles. With lysogeny, the viral genome is repressed and replicates in latent form as part of the bacterial genome.

template
In molecular biology, a structure that directs the formation of copies of itself, with particular reference to nucleic acid chains that direct the nucleotide sequence of newly polymerized chains by the rules of Watson-Crick base-pairing. In nucleic acids, the new strand is complementary, not identical, to its template.

terminal differentiation
Cell differentiation into a cell type that does not reproduce itself.

terminal repeat
A duplicated DNA sequence found at both ends of a defined genetic element, such as a transposon or LTR retrotransposon.

terminal transferase
An enzyme that adds nucleotides to the end of a nucleic acid chain in the absence of a template. Different terminal transferases operate on RNA and DNA.

terminator sequence
A DNA sequence that signals the end of transcription or replication.

tetracycline
An antibiotic that inhibits protein synthesis on the bacterial ribosome.

tetraploid
A cell or organism that has four copies of the genome.

thermophilic
Literally, "heat-loving." Applied to organisms that live at high temperatures.

TIR
Terminal inverted repeat. The symmetrical sequences that denote the boundaries of many DNA transposons.

tissue
An organized collection of specialized differentiated cells that has its own set of functions within a multicellular organism.

transcription (transcript)
The process of copying DNA into RNA (an RNA copy of a DNA segment).

transcription factor
A protein involved in executing or regulating transcription. Often (but not always) a DNA-binding protein.

transcription factory
A specialized subnuclear region in eukaryotic nuclei where transcription occurs. Typically characterized by only one of the three RNA polymerases.

transgenerational
Literally, "across generations." Applied to epigenetic modifications that are transmitted to one or more generations of progeny.

translation
In molecular biology, the process of converting the nucleotide sequence in messenger RNA into a sequence of amino acids in a polypeptide chain.

transposase
A protein that recognizes special transposon sequences and cleaves the donor site DNA at the transposon ends so that transposon DNA can insert into the target site.

transposon
A DNA segment that can move (transpose) from one location (donor site) to another in the genome (target site).

Trimethoprim
An antibiotic that inhibits bacterial tetrahydrofolate biosynthesis, essential for thymine monophosphate nucleotide production.

tRNA
Transfer RNA. One of a group of stable cell RNAs specific to individual amino acids that attach to and guide them to the right location in a polypeptide chain by recognizing the triplet code in messenger RNA molecules on the ribosome.

trypanosome
A unicellular flagellated eukaryotic parasite that moves between biting insect vectors and mammals, where it multiplies in the bloodstream and causes such maladies as sleeping sickness and Chagas disease.

tumor necrosis factor (TNF)
A cell-signaling molecule that triggers programmed cell death.

undulopodia
Literally, "waving foot." Cilia structures in a wide variety of eukaryotic cells with motility or sensory functions.

uniformitarianism
From Lyell's theories of geology, the concept that the same kinds of events cause small and large changes. Conceptually related to gradualism.

unique DNA sequence
A DNA sequence not repeated in the genome.

untemplated
In molecular biology, RNA or DNA chains synthesized by a terminal transferase without a template sequence from the genome.

variable region
The part of an antibody protein chain or its coding sequence that differs from one antibody to another.

V(D)J joining
In immunogenetics, the process of connecting DNA segments designated as variable, diversity, and joining cassettes to assemble complete loci encoding the two chains of antibody proteins.

vector
In biology, a carrier or transporter. In infectious disease, refers to an organism that transmits pathogens (such as mosquitoes that spread malarial parasites). In molecular genetics, refers to a DNA molecule that can be used to introduce a specific construct into a genetically modified organism. In genetics and evolution, refers to a virus or cell that carries genetic information from one organism to another.

vesicle
An extra- or intracellular structure enclosed by a membrane.

virus
An infectious agent composed of nucleic acid, proteins, and (sometimes) lipids that can reproduce only by infecting a host cell, where its genome encodes the synthesis and assembly of new virus components.

VNTR
Variable nucleotide tandem repeat. A form of repetitive DNA composed of variable numbers of short repeat sequences.

WGD
Whole genome duplication.

Wnt
A cell-cell signaling molecule found in animals. Named for its involvement in *Drosophila* wing development and in induction of mouse tumors following retroviral integration into the genome.

zygote
The diploid cell formed by fusion of two haploid gametes that goes on to divide and develop into an embryo.

References

1. Darwin, C. *Origin of Species* (John Russel, London, 1859).

2. Sniegowski, P.D. and Lenski, R.E. Mutation and adaptation: The directed mutation controversy in evolutionary perspective. *Annu Rev Ecol Systematics* 26, 553-578 (1995).

3. Merlin, F. Evolutionary Chance Mutation: A Defense Of The Modern Synthesis' Consensus View. *Philos Theor Biol* 2, e103 (2010).

4. McClintock, B. *Discovery And Characterization of Transposable Elements: The Collected Papers of Barbara McClintock* (Garland, New York, 1987).

5. Cohen, S.N. and Shapiro, J.A. Transposable genetic elements. *Sci Am* 242, 40-9 (1980).

6. Bukhari, A.I., J.A. Shapiro, and S. L. Adhya (Eds.). *DNA insertion elements, plasmids and episomes* (Cold Spring Harbor Press, Cold Spring Harbor, New York, 1977).

7. Shapiro, J.A. *Mobile Genetic Elements*, (Academic Press, New York, 1983).

8. Craig, N., Craigie, R, Gellert, M, Lambowitz, AM. *Mobile DNA II* (American Society for Microbiology Press, Washington, 2002).

9. Holliday, R. A Different Kind of Inheritance. *Sci Am* 260, 60-73 (1989).

10. Tan, S.Y. and Brown, J. Rudolph Virchow (1821-1902): "Pope of pathology." *Singapore Med J* 47, 567-8 (2006).

11. Wilson, P.G. Centriole inheritance. *Prion* 2, 9-16 (2008).

12. Harold, F.M. Molecules into cells: specifying spatial architecture. *Microbiol Mol Biol Rev* 69, 544-64 (2005).

13. McLaren, A. and Michie, D. Factors affecting vertebral variation in mice. 4. Experimental proof of the uterine basis of a maternal effect. *J Embryol Exp Morphol* 6, 645-59 (1958).

14. McLaren, A. and Michie, D. An effect of the uterine environment upon skeletal morphology in the mouse. *Nature* 181, 1147-8 (1958).

15. Grindstaff, J.L., Brodie, E.D., 3rd and Ketterson, E.D. Immune function across generations: integrating mechanism and evolutionary process in maternal antibody transmission. *Proc Biol Sci* 270, 2309-19 (2003).

16. Hasselquist, D. and Nilsson, J.A. Maternal transfer of antibodies in vertebrates: trans-generational effects on offspring immunity. *Philos Trans R Soc Lond B Biol Sci* 364, 51-60 (2009).

17. Sathananthan, A.H. Human centriole: origin, and how it impacts fertilization, embryogenesis, infertility and cloning. *Indian J Med Res* 129, 348-50 (2009).

18. Barroso, G. et al. Developmental sperm contributions: fertilization and beyond. *Fertil Steril* 92, 835-48 (2009).

19. Sonneborn, T.M. Partner of the Genes. *Sci Am* 183, 30-39 (1950).

20. Sonneborn, T.M. The determination of hereditary antigenic differences in genically identical Paramecium cells. *Proc Natl Acad Sci USA* 34, 413-8 (1948).

21. Sonneborn, T.M. Genetics of cell-cell interactions in ciliates. *Birth Defects Orig Artic Ser* 14, 417-27 (1978).

22. Beisson, J. Preformed cell structure and cell heredity. *Prion* 2, 1-8 (2008).

23. Prusiner, S.B. Prions. *Sci Am* 251, 50-59 (1984).

24. Sindi, S.S. and Serio, T.R. Prion dynamics and the quest for the genetic determinant in protein-only inheritance. *Curr Opin Microbiol* 12, 623-30 (2009).

25. Wickner, R.B. et al. Prion amyloid structure explains templating: how proteins can be genes. *FEMS Yeast Res* (2010).

26. Halfmann, R. and Lindquist, S. Epigenetics in the Extreme: Prions and the Inheritance of Environmentally Acquired Traits, *Science*. 330, 629-632 (2010).

27. Benkemoun, L. and Saupe, S.J. Prion proteins as genetic material in fungi. *Fungal Genet Biol* 43, 789-803 (2006).

28. Roberts, B.T. and Wickner, R.B. Heritable activity: a prion that propagates by covalent autoactivation. *Genes Dev* 17, 2083-7 (2003).

29. Tuite, M.F. and Serio, T.R. The prion hypothesis: from biological anomaly to basic regulatory mechanism. *Nat Rev Mol Cell Biol* 11, 823-33 (2010).

30. Alberts, B. et al. *Molecular Biology of the Cell* (Garland Science, New York and London, 2002).

31. Kuhn, T.S. *The Structure of Scientific Revolutions* (Univ. of Chicago Press, Chicago, 1962).

32. Simon, H.A. The Architecture of Complexity. *Proceedings of the American Philosophical Society* 106, 467-482 (1962).

33. Bhardwaj, N., Kim, P.M. and Gerstein, M.B. Rewiring of transcriptional regulatory networks: hierarchy, rather than connectivity, better reflects the importance of regulators. *Sci Signal* 3, ra79 (2010).

34. Welch, W.J. How Cells Respond to Stress. *Sci Am* 268, 56-64 (1993).

35. Duke, R.C., Ojcius, D.M. and Young, J.D. Cell Suicide in Health and Disease. *Sci Am* 275, 80-87 (1996).

36. Horvitz, H.R., Shaham, S. and Hengartner, M.O. The genetics of programmed cell death in the nematode Caenorhabditis elegans. *Cold Spring Harb Symp Quant Biol* 59, 377-85 (1994).

37. Pontier, D., Balague, C. and Roby, D. The hypersensitive response. A programmed cell death associated with plant resistance. *C R Acad Sci III* 321, 721-34 (1998).

38. Engelberg-Kulka, H., Amitai, S., Kolodkin-Gal, I. and Hazan, R. Bacterial programmed cell death and multicellular behavior in bacteria. *PLoS Genet* 2, e135 (2006).

39. Deponte, M. Programmed cell death in protists. *Biochim Biophys Acta* 1783, 1396-405 (2008).

40. Shapiro, J.A. Bacteria are small but not stupid: cognition, natural genetic engineering and sociobacteriology. *Stud Hist Philos Biol Biomed Sci* 38, 807-19 (2007).

41. Spiro, S., Dixon, R. and (eds.). *Sensory Mechanisms in Bacteria: Molecular Aspects of Signal Recognition* (Caister Academic Press, 2010).

42. Bray, D. *Wetware A Computer in Every Living Cell* (Yale University Press, New Haven, CT, 2009).

43. Danchin, A. Bacteria as computers making computers. *FEMS Microbiol Rev* 33, 3-26 (2009).

44. Kitano, H. Systems biology: a brief overview. *Science* 295, 1662-4 (2002).

45. Strange, K. The end of "naive reductionism": rise of systems biology or renaissance of physiology? *Am J Physiol Cell Physiol* 288, C968-74 (2005).

46. Noble, D. Biophysics and systems biology. *Philos Transact A Math Phys Eng Sci* 368, 1125-39 (2010).

47. Monod, J. *Recherches Sur La Croissance Des Cultures Bactériennes* (Hermann and cie, Paris, 1942).

48. Monod, J. From enzymatic adaptation to allosteric transitions. *Science* 154, 475-83 (1966).

49. Ullmann, A. Jacques Monod, 1910-1976: his life, his work and his commitments. *Res Microbiol* 161, 68-73 (2010).

50. Jacob, F. and Wollman, E.L. Viruses and Genes. *Sci Am* 204, 92-110 (1961).

51. Ptashne, M. and Gilbert, W. Genetic Repressors. *Sci Am* 222, 36-44 (1970).

52. Mueller-Hill, B. *The Lac Operon: A Short History of a Genetic Paradigm*, (de Gruyter, Berlin, 1996).

53. Jacob, F., Perrin, D., Sanchez, C. and Monod, J. Operon: a group of genes with the expression coordinated by an operator. *C. R. Hebd Seances Acad Sci* 250, 1727-1729 (1960).

54. Nathanson, J.A. and Greengard, P. "Second Messengers" in the Brain. *Sci Am* 237, 108-119 (1977).

55. Rasmussen, H. The Cycling of Calcium as an Intracellular Messenger. *Sci Am* 261, 66-73 (1989).

56. Scott, J.D. and Pawson, T. Cell Communication: The Inside Story. *Sci Am* 282, 72-79 (2000).

57. Raju, T.N. The Nobel chronicles. 1971: Earl Wilbur Sutherland, Jr. (1915-74). *Lancet* 354, 961 (1999).

58. Collado-Vides, J. A linguistic representation of the regulation of transcription initiation. II. Distinctive features of sigma 70 promoters and their regulatory binding sites. *Biosystems* 29, 105-28 (1993).

59. Barbieri, M. Biosemiotics: a new understanding of life. *Naturwissenschaften* 95, 577-99 (2008).

60. Jacob, F. and Monod, J. Genetic regulatory mechanisms in the synthesis of proteins. *J Mol Biol* 3, 318-56 (1961).

61. Lodish, H.F. et al. *Molecular Cell Biology* (W. H. Freeman and Co, New York, 1999).

62. Shapiro, J.A. and Sternberg, R.v. Why repetitive DNA is essential to genome function. *Biol. Revs. (Camb.)* 80, 227-50 (2005).

63. Istrail, S., De-Leon, S.B. and Davidson, E.H. The regulatory genome and the computer. *Dev Biol* 310, 187-95 (2007).

64. Tjian, R. Molecular Machines that Control Genes. *Sci Am* 272, 54-61 (1995).

65. Kuhlman, T., Zhang, Z., Saier, M.H., Jr. and Hwa, T. Combinatorial transcriptional control of the lactose operon of Escherichia coli. *Proc Natl Acad Sci USA* 104, 6043-8 (2007).

66. Muller, J., Barker, A., Oehler, S. and Muller-Hill, B. Dimeric lac repressors exhibit phase-dependent co-operativity. *J Mol Biol* 284, 851-7 (1998).

67. Muller, J., Oehler, S. and Muller-Hill, B. Repression of lac promoter as a function of distance, phase and quality of an auxiliary lac operator. *J Mol Biol* 257, 21-9 (1996).

68. Davidson, E.H. *The Regulatory Genome* (Academic, San Diego, 2006).

69. Changeux, J.-P. The Control of Biochemical Reactions. *Sci Am* 212, 36-45 (1965).

70. Koshland, D.E. Protein Shape and Biological Control. *Sci Am* 229, 52-64 (1973).

71. Monod, J., Changeux, J.P. and Jacob, F. Allosteric proteins and cellular control systems. *J Mol Biol* 6, 306-29 (1963).

72. Monod, J., Wyman, J. and Changeux, J.P. On the Nature of Allosteric Transitions: a Plausible Model. *J Mol Biol* 12, 88-118 (1965).

73. Kimata, K., Takahashi, H., Inada, T., Postma, P. and Aiba, H. cAMP receptor protein-cAMP plays a crucial role in glucose-lactose diauxie by activating the major glucose transporter gene in Escherichia coli. *Proc Natl Acad Sci USA* 94, 12914-9 (1997).

74. Lengeler, J.W. and Jahreis, K. Bacterial PEP-dependent carbohydrate: phosphotransferase systems couple sensing and global control mechanisms. *Contrib Microbiol* 16, 65-87 (2009).

75. Britten, R.J. and Davidson, E.H. Repetitive and non-repetitive DNA sequences and a speculation on the origins of evolutionary novelty. *Q Rev Biol* 46, 111-38 (1971).

76. Ravasz, E. Detecting hierarchical modularity in biological networks. *Methods Mol Biol* 541, 145-60 (2009).

77. Cooper, S. and Helmstetter, C.E. Chromosome replication and the division cycle of Escherichia coli B/r. *J Mol Biol* 31, 519-40 (1968).

78. Kunkel, T.A. and Bebenek, K. DNA replication fidelity. *Annu Rev Biochem* 69, 497-529 (2000).

79. Radman, M. and Wagner, R. The high fidelity of DNA duplication. *Sci Am* 259, 40-6 (1988).

80. Rennie, J. Proofreading Genes. *Sci Am* 264, 28-32 (1991).

81. Fazlieva, R. et al. Proofreading exonuclease activity of human DNA polymerase delta and its effects on lesion-bypass DNA synthesis. *Nucleic Acids Res* 37, 2854-66 (2009).

82. Ibarra, B. et al. Proofreading dynamics of a processive DNA polymerase. *Embo J* 28, 2794-802 (2009).

83. Modrich, P. and Lahue, R. Mismatch repair in replication fidelity, genetic recombination, and cancer biology. *Annu Rev Biochem* 65, 101-33 (1996).

84. Jiricny, J. The multifaceted mismatch-repair system. *Nat Rev Mol Cell Biol* 7, 335-46 (2006).

85. Iyer, R.R., Pluciennik, A., Burdett, V. and Modrich, P.L. DNA mismatch repair: functions and mechanisms. *Chem Rev* 106, 302-23 (2006).

86. Kunkel, T.A. and Erie, D.A. DNA mismatch repair. *Annu Rev Biochem* 74, 681-710 (2005).

87. Sancar, A. and Hearst, J.E. Molecular matchmakers. *Science* 259, 1415-20 (1993).

88. Bebenek, K. and Kunkel, T.A. Functions of DNA polymerases. *Adv Protein Chem* 69, 137-65 (2004).

89. Fujii, S. and Fuchs, R.P. Interplay among replicative and specialized DNA polymerases determines failure or success of translesion synthesis pathways. *J Mol Biol* 372, 883-93 (2007).

90. Garcia-Diaz, M. and Bebenek, K. Multiple functions of DNA polymerases. *CRC Crit Rev Plant Sci* 26, 105-122 (2007).

91. Nick McElhinny, S.A., Gordenin, D.A., Stith, C.M., Burgers, P.M. and Kunkel, T.A. Division of labor at the eukaryotic replication fork. *Mol Cell* 30, 137-44 (2008).

92. Peltomaki, P. and de la Chapelle, A. Mutations predisposing to hereditary nonpolyposis colorectal cancer. *Adv Cancer Res* 71, 93-119 (1997).

93. Kosko, B. and Isaka, S. Fuzzy Logic. *Sci Am* 269, 76-81 (1993).

94. Sadegh-Zadeh, K. Advances in fuzzy theory. *Artif Intell Med* 15, 309-323 (1999).

95. Jamshidi, M. Tools for intelligent control: fuzzy controllers, neural networks and genetic algorithms. *Phil Trans A: Math Phys Eng Sci* 361, 1781-1808 (2003).

96. Ganesan, A.K. and Smith, K.C. Dark recovery processes in Escherichia coli irradiated with ultraviolet light. I. Effect of rec mutations on liquid holding recovery. *J Bacteriol* 96, 365-73 (1968).

97. Das, J., Bagchi, B. and Chaudhuri, U. Liquid holding recovery of photodynamic damage in E. coli. *Photochem Photobiol* 19, 317-9 (1974).

98. Howard-Flanders, P. Inducible Repair of DNA. *Sci Am* 245, 72-80 (1981).

99. Weigle, J.J. Induction of Mutations in a Bacterial Virus. *Proc Natl Acad Sci USA* 39, 628-36 (1953).

100. Wood, R.D. and Hutchinson, F. Non-targeted mutagenesis of unirradiated lambda phage in Escherichia coli host cells irradiated with ultraviolet light. *J Mol Biol* 173, 293-305 (1984).

101. Maenhaut-Michel, G. Mechanism of SOS-induced targeted and untargeted mutagenesis in E. coli. *Biochimie* 67, 365-9 (1985).

102. Calsou, P. and Defais, M. Weigle reactivation and mutagenesis of bacteriophage lambda in lexA(Def) mutants of E. coli K12. *Mol Gen Genet* 201, 329-33 (1985).

103. Hutchinson, F. and Stein, J. Mutagenesis of ultraviolet-irradiated lambda phage by host cell irradiation: induction of Weigle mutagenesis is not an all-or-none process. *Mol Gen Genet* 177, 207-11 (1980).

104. Hutchinson, F. and Stein, J. Mutagenesis of lambda phage: Weigle mutagenesis is induced by coincident lesions in the double helical DNA of the host cell genome. *Mol Gen Genet* 181, 458-63 (1981).

105. Witkin, E.M. Elevated mutability of polA derivatives of Escherichia coli B/r at sublethal doses of ultraviolet light: evidence for an inducible error-prone repair system ("SOS repair") and its anomalous expression in these strains. *Genetics* 79, Suppl:199-213 (1975).

106. Battista, J.R., Donnelly, C.E., Ohta, T. and Walker, G.C. The SOS response and induced mutagenesis. *Prog Clin Biol Res* 340A, 169-78 (1990).

107. Janion, C. Inducible SOS response system of DNA repair and mutagenesis in Escherichia coli. *Int J Biol Sci* 4, 338-44 (2008).

108. Smith, K.C. Multiple pathways of DNA repair and their possible roles in mutagenesis. *Natl Cancer Inst Monogr*, 107-14 (1978).

109. Van Houten, B. Nucleotide excision repair in Escherichia coli. *Microbiol Rev* 54, 18-51 (1990).

110. Petit, C. and Sancar, A. Nucleotide excision repair: from E. coli to man. *Biochimie* 81, 15-25 (1999).

111. Pham, P., Rangarajan, S., Woodgate, R. and Goodman, M.F. Roles of DNA polymerases V and II in SOS-induced error-prone and error-free repair in Escherichia coli. *Proc Natl Acad Sci USA* 98, 8350-4 (2001).

112. Wagner, J., Etienne, H., Janel-Bintz, R. and Fuchs, R.P. Genetics of mutagenesis in E. coli: various combinations of translesion polymerases (Pol II, IV and V) deal with lesion/sequence context diversity. *DNA Repair (Amst)* 1, 159-67 (2002).

113. Goodman, M.F. Error-prone repair DNA polymerases in prokaryotes and eukaryotes. *Annu Rev Biochem* 71, 17-50 (2002).

114. Broyde, S., Wang, L., Rechkoblit, O., Geacintov, N.E. and Patel, D.J. Lesion processing: high-fidelity versus lesion-bypass DNA polymerases. *Trends Biochem Sci* 33, 209-19 (2008).

115. Napolitano, R., Janel-Bintz, R., Wagner, J. and Fuchs, R.P. All three SOS-inducible DNA polymerases (Pol II, Pol IV and Pol V) are involved in induced mutagenesis. *Embo J* 19, 6259-65 (2000).

116. Walker, G.C. Understanding the complexity of an organism's responses to DNA damage. *Cold Spring Harb Symp Quant Biol* 65, 1-10 (2000).

117. Sutton, M.D., Smith, B.T., Godoy, V.G. and Walker, G.C. The SOS response: recent insights into umuDC-dependent mutagenesis and DNA damage tolerance. *Annu Rev Genet* 34, 479-497 (2000).

118. Stahl, F.W. Genetic Recombination. *Sci Am* 256, 90-101 (1987).

119. Witkin, E.M. RecA protein in the SOS response: milestones and mysteries. *Biochimie* 73, 133-41 (1991).

120. Kowalczykowski, S.C., Dixon, D.A., Eggleston, A.K., Lauder, S.D. and Rehrauer, W.M. Biochemistry of homologous recombination in Escherichia coli. *Microbiol Rev* 58, 401-65 (1994).

121. Kowalczykowski, S.C. Initiation of genetic recombination and recombination-dependent replication. *Trends Biochem Sci* 25, 156-65 (2000).

122. Lusetti, S.L. and Cox, M.M. The bacterial RecA protein and the recombinational DNA repair of stalled replication forks. *Annu Rev Biochem* 71, 71-100 (2002).

123. Cox, M.M. Recombinational DNA repair of damaged replication forks in Escherichia coli: questions. *Annu Rev Genet* 35, 53-82 (2001).

124. Kidane, D. and Graumann, P.L. Dynamic formation of RecA filaments at DNA double strand break repair centers in live cells. *J Cell Biol* 170, 357-66 (2005).

125. Chen, Z., Yang, H. and Pavletich, N.P. Mechanism of homologous recombination from the RecA-ssDNA/dsDNA structures. *Nature* 453, 489-4 (2008).

126. Stasiak, A.Z., Rosselli, W. and Stasiak, A. RecA-DNA helical filaments in genetic recombination. *Biochimie* 73, 199-208 (1991).

127. Takahashi, M. and Norden, B. Structure of RecA-DNA complex and mechanism of DNA strand exchange reaction in homologous recombination. *Adv Biophys* 30, 1-35 (1994).

128. Roberts, J.W. and Roberts, C.W. Proteolytic cleavage of bacteriophage lambda repressor in induction. *Proc Natl Acad Sci USA* 72, 147-51 (1975).

129. Roberts, J.W., Roberts, C.W. and Craig, N.L. Escherichia coli recA gene product inactivates phage lambda repressor. *Proc Natl Acad Sci USA* 75, 4714-8 (1978).

130. Roberts, J.W., Phizicky, E.M., Burbee, D.G., Roberts, C.W. and Moreau, P.L. A brief consideration of the SOS inducing signal. *Biochimie* 64, 805-7 (1982).

131. Phizicky, E.M. and Roberts, J.W. Induction of SOS functions: regulation of proteolytic activity of E. coli RecA protein by interaction with DNA and nucleoside triphosphate. *Cell* 25, 259-67 (1981).

132. Craig, N.L. and Roberts, J.W. E. coli recA protein-directed cleavage of phage lambda repressor requires polynucleotide. *Nature* 283, 26-30 (1980).

133. McCann, J., Spingarn, N.E., Kobori, J. and Ames, B.N. Detection of carcinogens as mutagens: bacterial tester strains with R factor plasmids. *Proc Natl Acad Sci USA* 72, 979-83 (1975).

134. Inman, M.A., Butler, M.A., Connor, T.H. and Matney, T.S. The effects of excision repair and the plasmid pKM101 on the induction of his+ revertants by chemical agents in Salmonella typhimurium. *Teratog Carcinog Mutagen* 3, 491-501 (1983).

135. Walker, G.C. Mutagenesis-enhancement by plasmids in mutagenesis tester strains. *Basic Life Sci* 34, 111-20 (1985).

136. Devoret, R. Bacterial tests for potential carcinogens. *Sci Am* 241, 40-9 (1979).

137. Huisman, O., D'Ari, R. and Gottesman, S. Cell-division control in Escherichia coli: specific induction of the SOS function SfiA protein is sufficient to block septation. *Proc Natl Acad Sci USA* 81, 4490-4 (1984).

138. Higashitani, A., Higashitani, N. and Horiuchi, K. A cell division inhibitor SulA of Escherichia coli directly interacts with FtsZ through GTP hydrolysis. *Biochem Biophys Res Commun* 209, 198-204 (1995).

139. Hartwell, L. and Weinert, TA. Checkpoints: controls that ensure the order of cell cycle events. *Science* 246, 629-634 (1989).

140. Weinert, T.A. and Hartwell, L.H. The RAD9 gene controls the cell cycle response to DNA damage in Saccharomyces cerevisiae. *Science* 241, 317-22 (1988).

141. Watanabe, K., Morishita, J., Umezu, K., Shirahige, K. and Maki, H. Involvement of RAD9-dependent damage checkpoint control in arrest of cell cycle, induction of cell death, and chromosome instability caused by defects in origin recognition complex in Saccharomyces cerevisiae. *Eukaryot Cell* 1, 200-12 (2002).

142. Mazia, D. The Cell Cycle. *Sci Am* 230, 54-64 (1974).

143. Murray, A. and Kirschner, M. What controls the cell cycle. *Sci Am* 264, 56-63 (1991).

144. Weinberg, R.A. How cancer arises. *Sci Am* 275, 62-70 (1996).

145. Hartwell, L. Defects in a cell cycle checkpoint may be responsible for the genomic instability of cancer cells. *Cell* 71, 543-6 (1992).

146. Paulovich, A.G. and Hartwell, L.H. A checkpoint regulates the rate of progression through S phase in S. cerevisiae in response to DNA damage. *Cell* 82, 841-7 (1995).

147. Ishikawa, K., Ishii, H. and Saito, T. DNA damage-dependent cell cycle checkpoints and genomic stability. *DNA Cell Biol* 25, 406-11 (2006).

148. McIntosh, J.R. and McDonald, K.L. The Mitotic Spindle. *Sci Am* 261, 48-56 (1989).

149. Nicklas, R.B. How cells get the right chromosomes. *Science* 275, 632-7 (1997).

150. Hoyt, M.A. A new view of the spindle checkpoint. *J Cell Biol* 154, 909-11 (2001).

151. Taylor, S.S., Scott, M.I. and Holland, A.J. The spindle checkpoint: a quality control mechanism which ensures accurate chromosome segregation. *Chromosome Res* 12, 599-616 (2004).

152. Glover, D.M., Gonzalez, C. and Raff, J.W. The Centrosome. *Sci Am* 268, 62-68 (1993).

153. Santaguida, S. and Musacchio, A. The life and miracles of kinetochores. *Embo J* 28, 2511-31 (2009).

154. Burke, D.J. and Stukenberg, P.T. Linking kinetochore-microtubule binding to the spindle checkpoint. *Dev Cell* 14, 474-9 (2008).

155. Cross, F., Hartwell, L.H., Jackson, C. and Konopka, J.B. Conjugation in Saccharomyces cerevisiae. *Annu Rev Cell Biol* 4, 429-57 (1988).

156. Jackson, C.L. and Hartwell, L.H. Courtship in Saccharomyces cerevisiae: an early cell-cell interaction during mating. *Mol Cell Biol* 10, 2202-13 (1990).

157. Jackson, C.L., Konopka, J.B. and Hartwell, L.H. S. cerevisiae alpha pheromone receptors activate a novel signal transduction pathway for mating partner discrimination. *Cell* 67, 389-402 (1991).

158. Dorer, R., Pryciak, P., Schrick, K. and Hartwell, L.H. The induction of cell polarity by pheromone in Saccharomyces cerevisiae. *Harvey Lect* 90, 95-104 (1994).

159. Jackson, C.L. and Hartwell, L.H. Courtship in S. cerevisiae: both cell types choose mating partners by responding to the strongest pheromone signal. *Cell* 63, 1039-51 (1990).

160. Under, M.E. and Gilman, A.G. G Proteins. *Sci Am* 267, 56-65 (1992).

161. Ameisen, J.C. On the origin, evolution, and nature of programmed cell death: a timeline of four billion years. *Cell Death Differ* 9, 367-93 (2002).

162. Huettenbrenner, S. et al. The evolution of cell death programs as prerequisites of multicellularity. *Mutat Res* 543, 235-49 (2003).

163. Cheng, W.C., Leach, K.M. and Hardwick, J.M. Mitochondrial death pathways in yeast and mammalian cells. *Biochim Biophys Acta* 1783, 1272-9 (2008).

164. Zuzarte-Luis, V. and Hurle, J.M. Programmed cell death in the embryonic vertebrate limb. *Semin Cell Dev Biol* 16, 261-9 (2005).

165. Burhans, W.C. et al. Apoptosis-like yeast cell death in response to DNA damage and replication defects. *Mutat Res* 532, 227-43 (2003).

166. Fulda, S., Gorman, A.M., Hori, O. and Samali, A. Cellular stress responses: cell survival and cell death. *Int J Cell Biol*, 214-74 (2010).

167. Heath, M.C. Hypersensitive response-related death. *Plant Mol Biol* 44, 321-34 (2000).

168. Kolodkin-Gal, I., Hazan, R., Gaathon, A., Carmeli, S. and Engelberg-Kulka, H. A linear pentapeptide is a quorum-sensing factor required for mazEF-mediated cell death in Escherichia coli. *Science* 318, 652-5 (2007).

169. Kolodkin-Gal, I. and Engelberg-Kulka, H. The extracellular death factor: physiological and genetic factors influencing its production and response in Escherichia coli. *J Bacteriol* 190, 3169-75 (2008).

170. Jarpe, M.B. et al. Anti-apoptotic versus pro-apoptotic signal transduction: checkpoints and stop signs along the road to death. *Oncogene* 17, 1475-82 (1998).

171. Marini, P. and Belka, C. Death receptor ligands: new strategies for combined treatment with ionizing radiation. *Curr Med Chem Anticancer Agents* 3, 334-42 (2003).

172. Holoch, P.A. and Griffith, T.S. TNF-related apoptosis-inducing ligand (TRAIL): a new path to anti-cancer therapies. *Eur J Pharmacol* 625, 63-72 (2009).

173. Linseman, D.A. et al. Insulin-like growth factor-I blocks Bcl-2 interacting mediator of cell death (Bim) induction and intrinsic death signaling in cerebellar granule neurons. *J Neurosci* 22, 9287-97 (2002).

174. Linseman, D.A. et al. Suppression of death receptor signaling in cerebellar Purkinje neurons protects neighboring granule neurons from apoptosis via an insulin-like growth factor I-dependent mechanism. *J Biol Chem* 277, 24546-53 (2002).

175. Slee, E.A. and Lu, X. The ASPP family: deciding between life and death after DNA damage. *Toxicol Lett* 139, 81-7 (2003).

176. Loewer, A. and Lahav, G. Cellular conference call: external feedback affects cell-fate decisions. *Cell* 124, 1128-30 (2006).

177. Song, J. EMT or apoptosis: a decision for TGF-beta. *Cell Res* 17, 289-90 (2007).

178. Falschlehner, C., Emmerich, C.H., Gerlach, B. and Walczak, H. TRAIL signalling: decisions between life and death. *Int J Biochem Cell Biol* 39, 1462-75 (2007).

179. Meredith, J.E., Jr., Fazeli, B. and Schwartz, M.A. The extracellular matrix as a cell survival factor. *Mol Biol Cell* 4, 953-61 (1993).

180. Park, D.S. et al. Multiple pathways of neuronal death induced by DNA-damaging agents, NGF deprivation, and oxidative stress. *J Neurosci* 18, 830-40 (1998).

181. O'Rourke, D.M. et al. Conversion of a radioresistant phenotype to a more sensitive one by disabling erbB receptor signaling in human cancer cells. *Proc Natl Acad Sci USA* 95, 10842-7 (1998).

182. Crick, F. On protein synthesis. *Symp Soc Exp Biol* 12, 138-163 (1958).

183. Temin, H.M. RNA-Directed DNA Synthesis. *Sci Am* 226, 24-33 (1972).

184. Temin, H. and S Mizutani. RNA-dependent DNA polymerase in virions of Rous sarcoma virus. *Nature* 226, 1211-1213 (1970).

185. Crick, F. Central dogma of molecular biology. *Nature* 227, 561-563 (1970).

186. Collins, F.S. and Jegalian, K.G. Deciphering the Code of Life. *Sci Am* 281, 86-91 (1999).

187. Shapiro, J.A. Genome informatics: The role of DNA in cellular computations. *Biological Theory* 1, 288-301 (2006).

188. Shapiro, J.A. Revisiting the Central Dogma in the 21st Century. *Annals of the New York Academy of Sciences* 1178, 6-28 (2009).

189. Judson, H. *The Eighth Day of Creation: Makers of the Revolution in Biology*, (Simon and Schuster, New York, 1979).

190. Huxley, J. *Evolution: The Modern Synthesis*, (Allen and Unwin, London, 1942).

191. Koonin, E.V. Darwinian evolution in the light of genomics. *Nucleic Acids Res* 37, 1011-34 (2009).

192. Kutschera, U. and Niklas, K.J. The modern theory of biological evolution: an expanded synthesis. *Naturwissenschaften* 91, 255-76 (2004).

193. Rose, M.R. and Oakley, T.H. The new biology: beyond the Modern Synthesis. *Biol Direct* 2, 30 (2007).

194. Ryan, F.P. Genomic creativity and natural selection: a modern synthesis. *Biol J Linnean Society* 88, 655-672 (2006).

195. Britten, R.J. Cases of ancient mobile element DNA insertions that now affect gene regulation. *Mol Phylogenet Evol* 5, 13-7 (1996).

196. Britten, R.J. DNA sequence insertion and evolutionary variation in gene regulation. *Proc Natl Acad Sci USA* 93, 9374-7 (1996).

197. Britten, R. and Kohne, DE. Repeated sequences in DNA. Hundreds of thousands of copies of DNA sequences have been incorporated into the genomes of higher organisms. *Science* 161, 529-540 (1968).

198. Shapiro, J.A. Natural genetic engineering in evolution. *Genetica* 86, 99-111 (1992).

199. Shapiro, J.A. Genome system architecture and natural genetic engineering in evolution. *Annu NY Acad Sci* 870, 23-35 (1999).

200. Shapiro, J.A. Transposable elements as the key to a 21st century view of evolution. *Genetica* 107, 171-9 (1999).

201. Shapiro, J.A. Retrotransposons and regulatory suites. *Bioessays* 27, 122-5 (2005).

202. Shapiro, J.A. A 21st century view of evolution: genome system architecture, repetitive DNA, and natural genetic engineering. *Gene* 345, 91-100 (2005).

203. Sternberg, R.v. and Shapiro, J.A. How repeated retroelements format genome function. *Cytogenet Genome Res* 110, 108-116 (2005).

204. Shapiro, J.A. Mobile DNA and evolution in the 21st century. *Mob DNA* 1, 4 (2010).

205. Long, M. Evolution of novel genes. *Curr Opin Genet Dev* 11, 673-80 (2001).

206. Betrán, E., Thornton, K. and Long, M. Retroposed new genes out of the X in Drosophila. *Genome Res* 12, 1854-1859 (2002).

207. Jorgensen, R.A. Restructuring the genome in response to adaptive challenge: McClintock's bold conjecture revisited. *Cold Spring Harb Symp Quant Biol* 69, 349-54 (2004).

208. Bourque, G. Transposable elements in gene regulation and in the evolution of vertebrate genomes. *Curr Opin Genet Dev* 19, 607-12 (2009).

209. Bourque, G. et al. Evolution of the mammalian transcription factor binding repertoire via transposable elements. *Genome Res* 18, 1752-62 (2008).

210. Marino-Ramirez, L., Lewis, K.C., Landsman, D. and Jordan, I.K. Transposable elements donate lineage-specific regulatory sequences to host genomes. *Cytogenet Genome Res* 110, 333-41 (2005).

211. Jordan, I.K., Rogozin, I.B., Glazko, G.V. and Koonin, E.V. Origin of a substantial fraction of human regulatory sequences from transposable elements. *Trends Genet* 19, 68-72 (2003).

212. Bowen, N.J. and Jordan, I.K. Transposable elements and the evolution of eukaryotic complexity. *Curr Issues Mol Biol* 4, 65-76 (2002).

213. Jiang, N., Bao, Z., Zhang, X., Eddy, S.R. and Wessler, S.R. Pack-MULE transposable elements mediate gene evolution in plants. *Nature* 431, 569-73 (2004).

214. Wessler, S.R., Bureau, T.E. and White, S.E. LTR-retrotransposons and MITEs: important players in the evolution of plant genomes. *Curr Opin Genet Dev* 5, 814-21 (1995).

215. Wessler, S.R. Transposable elements and the evolution of gene expression. *Symp Soc Exp Biol* 51, 115-22 (1998).

216. Wessler, S.R. Transposable elements and the evolution of eukaryotic genomes. *Proc Natl Acad Sci USA* 103, 17600-1 (2006).

217. Piriyapongsa, J., Marino-Ramirez, L. and Jordan, I.K. Origin and evolution of human microRNAs from transposable elements. *Genetics* 176, 1323-37 (2007).

218. Brandt, J. et al. Transposable elements as a source of genetic innovation: expression and evolution of a family of retrotransposon-derived neogenes in mammals. *Gene* 345, 101-11 (2005).

219. Volff, J.N. and Brosius, J. Modern genomes with retro-look: retrotransposed elements, retroposition and the origin of new genes. *Genome Dyn* 3, 175-90 (2007).

220. Oliver, K.R. and Greene, W.K. Transposable elements: powerful facilitators of evolution. *Bioessays* 31, 703-14 (2009).

221. Deininger, P.L., Moran, J.V., Batzer, M.A. and Kazazian, H.H., Jr. Mobile elements and mammalian genome evolution. *Curr Opin Genet Dev* 13, 651-8 (2003).

222. Kazazian, H.H., Jr. Mobile elements: drivers of genome evolution. *Science* 303, 1626-32 (2004).

223. Bohne, A., Brunet, F., Galiana-Arnoux, D., Schultheis, C. and Volff, J.N. Transposable elements as drivers of genomic and biological diversity in vertebrates. *Chromosome Res* 16, 203-15 (2008).

224. Biemont, C. and Vieira, C. What transposable elements tell us about genome organization and evolution: the case of Drosophila. *Cytogenet Genome Res* 110, 25-34 (2005).

225. Arguello, J.R., Fan, C., Wang, W. and Long, M. Origination of chimeric genes through DNA-level recombination. *Genome Dyn* 3, 131-46 (2007).

226. Feschotte, C. and Pritham, E.J. DNA transposons and the evolution of eukaryotic genomes. *Annu Rev Genet* 41, 331-68 (2007).

227. Beadle, G.W. The genes of men and molds. *Sci Am* 179, 30-9 (1948).

228. Ast, G. The Alternative Genome. *Sci Am* 292, 58-65 (2005).

229. Portin, P. The elusive concept of the gene. *Hereditas* 146, 112-7 (2009).

230. Hinman, V.F., Yankura, K.A. and McCauley, B.S. Evolution of gene regulatory network architectures: examples of subcircuit conservation and plasticity between classes of echinoderms. *Biochim Biophys Acta* 1789, 326-32 (2009).

231. Erwin, D.H. and Davidson, E.H. The evolution of hierarchical gene regulatory networks. *Nat Rev Genet* 10, 141-8 (2009).

232. Ray, P.S., Arif, A. and Fox, P.L. Macromolecular complexes as depots for releasable regulatory proteins. *Trends Biochem Sci* 32, 158-64 (2007).

233. McClintock, B. Intranuclear systems controlling gene action and mutation. *Brookhaven Symp Biol* 58-74 (1956).

234. Wunderlich, Z. and Mirny, L.A. Different gene regulation strategies revealed by analysis of binding motifs. *Trends Genet* 25, 434-40 (2009).

235. Ptashne, M. and Gann, A. *Genes and Signals* (Laboratory Press, Cold Spring Harbor, New York, 2002).

236. Waddington, C.H. How do Cells Differentiate? *Sci Am* 189, 108-116 (1953).

237. Waddington, C.H. and Robertson, E. Selection for developmental canalisation. *Genet Res* 7, 303-12 (1966).

238. Waddington, C.H. Gene regulation in higher cells. *Science* 166, 639-40 (1969).

239. Van Speybroeck, L. From epigenesis to epigenetics: the case of C. H. Waddington. *Annu NY Acad Sci* 981, 61-81 (2002).

240. Jablonka, E. and Lamb, M.J. The changing concept of epigenetics. *Annu NY Acad Sci* 981, 82-96 (2002).

241. Holliday, R. Epigenetics: a historical overview. *Epigenetics* 1, 76-80 (2006).

242. Reik, W., Dean, W. and Walter, J. Epigenetic reprogramming in mammalian development. *Science* 293, 1089-93 (2001).

243. Reik, W. Stability and flexibility of epigenetic gene regulation in mammalian development. *Nature* 447, 425-32 (2007).

244. Jaenisch, R., Hochedlinger, K. and Eggan, K. Nuclear cloning, epigenetic reprogramming and cellular differentiation. *Novartis Found Symp* 265, 107-18; discussion 118-28 (2005).

245. Feng, S., Jacobsen, S.E. and Reik, W. Epigenetic Reprogramming in Plant and Animal Development. *Science* 330, 622-627 (2010).

246. Sapienza, C. Parental imprinting of genes. *Sci Am* 263, 52-60 (1990).

247. Shire, J.G. Unequal parental contributions: genomic imprinting in mammals. *New Biol* 1, 115-20 (1989).

248. Sapienza, C. Genome imprinting: an overview. *Dev Genet* 17, 185-7 (1995).

249. Ohlsson, R., Tycko, B. and Sapienza, C. Monoallelic expression: "there can only be one." *Trends Genet* 14, 435-8 (1998).

250. Nur, U. Heterochromatization and euchromatization of whole genomes in scale insects (Coccoidea: Homoptera). *Dev Suppl*, 29-34 (1990).

251. van Driel, R., Fransz, P.F. and Verschure, P.J. The eukaryotic genome: a system regulated at different hierarchical levels. *J Cell Sci* 116, 4067-75 (2003).

252. Chandler, V.L. Paramutation: from maize to mice. *Cell* 128, 641-5 (2007).

253. Cuzin, F., Grandjean, V. and Rassoulzadegan, M. Inherited variation at the epigenetic level: paramutation from the plant to the mouse. *Curr Opin Genet Dev* 18, 193-6 (2008).

254. Chandler, V.L. Paramutation's Properties and Puzzles. *Science* 330, 628-629 (2010).

255. Weaver, I.C. Epigenetic effects of glucocorticoids. *Semin Fetal Neonatal Med* 14, 143-50 (2009).

256. Weaver, J.R., Susiarjo, M. and Bartolomei, M.S. Imprinting and epigenetic changes in the early embryo. *Mamm Genome* 20, 532-43 (2009).

257. Skinner, M.K., Manikkam, M. and Guerrero-Bosagna, C. Epigenetic transgenerational actions of environmental factors in disease etiology. *Trends Endocrinol Metab* 21, 214-22 (2010).

258. Anway, M.D. and Skinner, M.K. Transgenerational effects of the endocrine disruptor vinclozolin on the prostate transcriptome and adult onset disease. *Prostate* 68, 517-29 (2008).

259. Boyko, A. and Kovalchuk, I. Transgenerational response to stress in Arabidopsis thaliana. *Plant Signal Behav* 5 (2010).

260. Boyko, A. et al. Transgenerational adaptation of Arabidopsis to stress requires DNA methylation and the function of Dicer-like proteins. *PLoS One* 5, e9514 (2010).

261. Lang-Mladek, C. et al. Transgenerational inheritance and resetting of stress-induced loss of epigenetic gene silencing in Arabidopsis. *Mol Plant* 3, 594-602 (2010).

262. Szyf, M. The early life environment and the epigenome. *Biochim Biophys Acta* 1790, 878-85 (2009).

263. Lange, U.C. and Schneider, R. What an epigenome remembers. *Bioessays* 32, 659-68 (2010).

264. Henikoff, S. Heterochromatin function in complex genomes. *Biochim Biophys Acta* 1470, O1-8 (2000).

265. Elgin, S.C. and Grewal, S.I. Heterochromatin: silence is golden. *Curr Biol* 13, R895-8 (2003).

266. Huisinga, K.L., Brower-Toland, B. and Elgin, S.C. The contradictory definitions of heterochromatin: transcription and silencing. *Chromosoma* 115, 110-22 (2006).

267. Spradling, A.C. Position effect variegation and genomic instability. *Cold Spring Harb Symp Quant Biol* 58, 585-96 (1993).

268. Henikoff, S. A reconsideration of the mechanism of position effect. *Genetics* 138, 1-5 (1994).

269. Karpen, G.H. Position-effect variegation and the new biology of heterochromatin. *Curr Opin Genet Dev* 4, 281-91 (1994).

270. Hazelrigg, T., Levis, R. and Rubin, G.M. Transformation of white locus DNA in drosophila: dosage compensation, zeste interaction, and position effects. *Cell* 36, 469-81 (1984).

271. Kornberg, R.D. and Klug, A. The Nucleosome. *Sci Am* 244, 52-64 (1981).

272. Grunstein, M. Histones as Regulators of Genes. *Sci Am* 267, 68-74 (1992).

273. McBryant, S.J., Adams, V.H. and Hansen, J.C. Chromatin architectural proteins. *Chromosome Res* 14, 39-51 (2006).

274. Chodaparambil, J.V., Edayathumangalam, R.S., Bao, Y., Park, Y.J. and Luger, K. Nucleosome structure and function. *Ernst Schering Res Found Workshop*, 29-46 (2006).

275. Segal, E. et al. A genomic code for nucleosome positioning. *Nature* 442, 772-8 (2006).

276. Peckham, H.E. et al. Nucleosome positioning signals in genomic DNA. *Genome Res* 17, 1170-7 (2007).

277. Kaplan, N. et al. The DNA-encoded nucleosome organization of a eukaryotic genome. *Nature* 458, 362-6 (2009).

278. Sapienza, C., Peterson, A.C., Rossant, J. and Balling, R. Degree of methylation of transgenes is dependent on gamete of origin. *Nature* 328, 251-4 (1987).

279. Henikoff, S., Furuyama, T. and Ahmad, K. Histone variants, nucleosome assembly and epigenetic inheritance. *Trends Genet* 20, 320-6 (2004).

280. Chen, Z.J. and Tian, L. Roles of dynamic and reversible histone acetylation in plant development and polyploidy. *Biochim Biophys Acta* 1769, 295-307 (2007).

281. Jenuwein, T. and Allis, C.D. Translating the histone code. *Science* 293, 1074-80 (2001).

282. Peterson, C.L. and Laniel, M.A. Histones and histone modifications. *Curr Biol* 14, R546-51 (2004).

283. Sims, R.J., 3rd and Reinberg, D. Is there a code embedded in proteins that is based on post-translational modifications? *Nat Rev Mol Cell Biol* 9, 815-20 (2008).

284. Campos, E.I. and Reinberg, D. Histones: Annotating Chromatin. *Annu Rev Genet* 43, 559-599 (2009).

285. Margueron, R., Trojer, P. and Reinberg, D. The key to development: interpreting the histone code? *Curr Opin Genet Dev* 15, 163-76 (2005).

286. Appelgren, H., Kniola, B. and Ekwall, K. Distinct centromere domain structures with separate functions demonstrated in live fission yeast cells. *J Cell Sci* 116, 4035-42 (2003).

287. Pidoux, A.L. and Allshire, R.C. The role of heterochromatin in centromere function. *Philos Trans R Soc Lond B Biol Sci* 360, 569-79 (2005).

288. Scott, K.C., Merrett, S.L. and Willard, H.F. A heterochromatin barrier partitions the fission yeast centromere into discrete chromatin domains. *Curr Biol* 16, 119-29 (2006).

289. Bonisch, C., Nieratschker, S.M., Orfanos, N.K. and Hake, S.B. Chromatin proteomics and epigenetic regulatory circuits. *Expert Rev Proteomics* 5, 105-19 (2008).

290. Holmes, R. and Soloway, P.D. Regulation of imprinted DNA methylation. *Cytogenet Genome Res* 113, 122-9 (2006).

291. Zheng, Y.G., Wu, J., Chen, Z. and Goodman, M. Chemical regulation of epigenetic modifications: opportunities for new cancer therapy. *Med Res Rev* 28, 645-87 (2008).

292. Choudhary, C. et al. Lysine acetylation targets protein complexes and co-regulates major cellular functions. *Science* 325, 834-40 (2009).

293. Szyf, M. The dynamic epigenome and its implications in toxicology. *Toxicol Sci* 100, 7-23 (2007).

294. Cairns, B.R. The logic of chromatin architecture and remodelling at promoters. *Nature* 461, 193-8 (2009).

295. Ikegami, K., Ohgane, J., Tanaka, S., Yagi, S. and Shiota, K. Interplay between DNA methylation, histone modification and chromatin remodeling in stem cells and during development. *Int J Dev Biol* 53, 203-14 (2009).

296. Mattick, J.S. The Hidden Genetic Program of Complex Organisms. *Sci Am* 291, 60-67 (2004).

297. Kelley, R.L. and Kuroda, M.I. Noncoding RNA genes in dosage compensation and imprinting. *Cell* 103, 9-12 (2000).

298. Mattick, J.S. Non-coding RNAs: the architects of eukaryotic complexity. *EMBO Rep* 2, 986-91 (2001).

299. Matzke, M.A. and Birchler, J.A. RNAi-mediated pathways in the nucleus. *Nat Rev Genet* 6, 24-35 (2005).

300. Amaral, P.P. and Mattick, J.S. Noncoding RNA in development. *Mamm Genome* 19, 454-92 (2008).

301. Andersen, A.A. and Panning, B. Epigenetic gene regulation by noncoding RNAs. *Curr Opin Cell Biol* 15, 281-9 (2003).

302. Martienssen, R.A., Zaratiegui, M. and Goto, D.B. RNA interference and hetero-chromatin in the fission yeast Schizosaccharomyces pombe. *Trends Genet* 21, 450-6 (2005).

303. Verdel, A., Vavasseur, A., Le Gorrec, M. and Touat-Todeschini, L. Common themes in siRNA-mediated epigenetic silencing pathways. *Int J Dev Biol* 53, 245-57 (2009).

304. Koerner, M.V., Pauler, F.M., Huang, R. and Barlow, D.P. The function of non-coding RNAs in genomic imprinting. *Development* 136, 1771-83 (2009).

305. Wheeler, B.S., Blau, J.A., Willard, H.F. and Scott, K.C. The impact of local genome sequence on defining heterochromatin domains. *PLoS Genet* 5, e1000453 (2009).

306. Kellum, R. and Elgin, S.C. Chromatin boundaries: punctuating the genome. *Curr Biol* 8, R521-4 (1998).

307. Bushey, A.M., Dorman, E.R. and Corces, V.G. Chromatin insulators: regula-tory mechanisms and epigenetic inheritance. *Mol Cell* 32, 1-9 (2008).

308. Eissenberg, J.C. and Elgin, S.C. Boundary functions in the control of gene expression. *Trends Genet* 7, 335-40 (1991).

309. Kellum, R. and Schedl, P. A group of scs elements function as domain bound-aries in an enhancer-blocking assay. *Mol Cell Biol* 12, 2424-31 (1992).

310. Gerasimova, T.I., Byrd, K. and Corces, V.G. A chromatin insulator determines the nuclear localization of DNA. *Mol Cell* 6, 1025-35 (2000).

311. de Laat, W. and Grosveld, F. Spatial organization of gene expression: the active chromatin hub. *Chromosome Res* 11, 447-59 (2003).

312. Osborne, C.S. et al. Active genes dynamically colocalize to shared sites of ongoing transcription. *Nat Genet* 36, 1065-71 (2004).

313. Scott, K.C., White, C.V. and Willard, H.F. An RNA polymerase III-dependent heterochromatin barrier at fission yeast centromere 1. *PLoS One* 2, e1099 (2007).

314. Sutherland, H. and Bickmore, W.A. Transcription factories: gene expression in unions? *Nat Rev Genet* 10, 457-66 (2009).

315. Lippman, Z. et al. Role of transposable elements in heterochromatin and epi-genetic control. *Nature* 430, 471-6 (2004).

316. Kotnova, A.P., Salenko, V.B., Lyubomirskaya, N.V. and Ilyin, Y.V. Structural organization of heterochromatin in Drosophila melanogaster: inverted repeats of transposable element clusters. *Dokl Biochem Biophys* 429, 293-5 (2009).

317. Buhler, M. and Gasser, S.M. Silent chromatin at the middle and ends: lessons from yeasts. *Embo J* 28, 2149-2161 (2009).

318. Guetg, C. et al. The NoRC complex mediates the heterochromatin formation and stability of silent rRNA genes and centromeric repeats. *Embo J* 29 (2010).

319. Thomas, C.M. Paradigms of plasmid organization. *Mol Microbiol* 37, 485-91 (2000).

320. Contursi, P. et al. Identification and autonomous replication capability of a chromosomal replication origin from the archaeon Sulfolobus solfataricus. *Extremophiles* 8, 385-91 (2004).

321. O'Donnell, M. Replisome architecture and dynamics in Escherichia coli. *J Biol Chem* 281, 10653-6 (2006).

322. Toro, E. and Shapiro, L. Bacterial chromosome organization and segregation. *Cold Spring Harb Perspect Biol* 2, a000349 (2010).

323. Berbenetz, N.M., Nislow, C. and Brown, G.W. Diversity of Eukaryotic DNA Replication Origins Revealed by Genome-Wide Analysis of Chromatin Structure. *PLoS Genet* 6, e1001092 (2010).

324. Sclafani, R.A. and Holzen, T.M. Cell cycle regulation of DNA replication. *Annu Rev Genet* 41, 237-80 (2007).

325. Muck, J. and Zink, D. Nuclear organization and dynamics of DNA replication in eukaryotes. *Front Biosci* 14, 5361-71 (2009).

326. Hazan, R. and Ben-Yehuda, S. Resolving chromosome segregation in bacteria. *J Mol Microbiol Biotechnol* 11, 126-39 (2006).

327. Ebersbach, G. and Gerdes, K. Plasmid segregation mechanisms. *Annu Rev Genet* 39, 453-79 (2005).

328. Tourand, Y., Lee, L. and Chaconas, G. Telomere resolution by Borrelia burgdorferi ResT through the collaborative efforts of tethered DNA binding domains. *Mol Microbiol* 64, 580-90 (2007).

329. Kobryn, K. and Chaconas, G. The circle is broken: telomere resolution in linear replicons. *Curr Opin Microbiol* 4, 558-64 (2001).

330. Greider, C.W. and Blackburn, E.H. Telomeres, Telomerase and Cancer. *Sci Am* 274, 92-97 (1996).

331. Pardue, M.L. and DeBaryshe, P.G. Drosophila telomeres: two transposable elements with important roles in chromosomes. *Genetica* 107, 189-96 (1999).

332. Hug, N. and Lingner, J. Telomere length homeostasis. *Chromosoma* 115, 413-25 (2006).

333. Grandin, N. and Charbonneau, M. Protection against chromosome degradation at the telomeres. *Biochimie* 90, 41-59 (2008).

334. Villasante, A., de Pablos, B., Mendez-Lago, M. and Abad, J.P. Telomere maintenance in Drosophila: rapid transposon evolution at chromosome ends. *Cell Cycle* 7, 2134-8 (2008).

335. Misri, S., Pandita, S., Kumar, R. and Pandita, T.K. Telomeres, histone code, and DNA damage response. *Cytogenet Genome Res* 122, 297-307 (2008).

336. Schoeftner, S. and Blasco, M.A. A "higher order" of telomere regulation: telomere heterochromatin and telomeric RNAs. *Embo J* 28, 2323-36 (2009).

337. Moser, B.A. and Nakamura, T.M. Protection and replication of telomeres in fission yeast. *Biochem Cell Biol* 87, 747-58 (2009).

338. Lowe, J. and Amos, L.A. Evolution of cytomotive filaments: the cytoskeleton from prokaryotes to eukaryotes. *Int J Biochem Cell Biol* 41, 323-9 (2009).

339. Thanbichler, M. Synchronization of chromosome dynamics and cell division in bacteria. *Cold Spring Harb Perspect Biol* 2, a000331 (2010).

340. Schumacher, M.A. Structural biology of plasmid partition: uncovering the molecular mechanisms of DNA segregation. *Biochem J* 412, 1-18 (2008).

341. Sharp, J.A. and Kaufman, P.D. Chromatin proteins are determinants of centromere function. *Curr Top Microbiol Immunol* 274, 23-52 (2003).

342. Vos, L.J., Famulski, J.K. and Chan, G.K. How to build a centromere: from centromeric and pericentromeric chromatin to kinetochore assembly. *Biochem Cell Biol* 84, 619-39 (2006).

343. Heit, R., Underhill, D.A., Chan, G. and Hendzel, M.J. Epigenetic regulation of centromere formation and kinetochore function. *Biochem Cell Biol* 84, 605-18 (2006).

344. Wenner, M. Nuclear Architecture. *Sci Am* 301, 20-22 (2009).

345. Hozak, P. The nucleoskeleton and attached activities. *Exp Cell Res* 229, 267-71 (1996).

346. Dechat, T. et al. Nuclear lamins: major factors in the structural organization and function of the nucleus and chromatin. *Genes Dev* 22, 832-53 (2008).

347. Fedorova, E. and Zink, D. Nuclear genome organization: common themes and individual patterns. *Curr Opin Genet Dev* 19, 166-71 (2009).

348. Zhao, R., Bodnar, M.S. and Spector, D.L. Nuclear neighborhoods and gene expression. *Curr Opin Genet Dev* 19, 172-9 (2009).

349. O'Sullivan, J.M., Sontam, D.M., Grierson, R. and Jones, B. Repeated elements coordinate the spatial organization of the yeast genome. *Yeast* 26, 125-38 (2009).

350. Pombo, A. et al. Specialized transcription factories within mammalian nuclei. *Crit Rev Eukaryot Gene Expr* 10, 21-9 (2000).

351. Bartlett, J. et al. Specialized transcription factories. *Biochem Soc Symp* 67-75 (2006).

352. Xu, M. and Cook, P.R. The role of specialized transcription factories in chromosome pairing. *Biochim Biophys Acta* 1783, 2155-60 (2008).

353. Faro-Trindade, I. and Cook, P.R. Transcription factories: structures conserved during differentiation and evolution. *Biochem Soc Trans* 34, 1133-7 (2006).

354. Xu, M. and Cook, P.R. Similar active genes cluster in specialized transcription factories. *J Cell Biol* 181, 615-23 (2008).

355. Mitchell, J.A. and Fraser, P. Transcription factories are nuclear subcompartments that remain in the absence of transcription. *Genes Dev* 22, 20-5 (2008).

356. Sandqvist, A. and Sistonen, L. Nuclear stress granules: the awakening of a sleeping beauty? *J Cell Biol* 164, 15-7 (2004).

357. Chiodi, I. et al. RNA recognition motif 2 directs the recruitment of SF2/ASF to nuclear stress bodies. *Nucleic Acids Res* 32, 4127-36 (2004).

358. Brown, J.M. et al. Association between active genes occurs at nuclear speckles and is modulated by chromatin environment. *J Cell Biol* 182, 1083-97 (2008).

359. Schneider, R. and Grosschedl, R. Dynamics and interplay of nuclear architecture, genome organization, and gene expression. *Genes Dev* 21, 3027-43 (2007).

360. Martin, W. and Koonin, E.V. Introns and the origin of nucleus-cytosol compartmentalization. *Nature* 440, 41-5 (2006).

361. Haaf, T., Golub, E.I., Reddy, G., Radding, C.M. and Ward, D.C. Nuclear foci of mammalian Rad51 recombination protein in somatic cells after DNA damage and its localization in synaptonemal complexes. *Proc Natl Acad Sci USA* 92, 2298-302 (1995).

362. Raderschall, E., Golub, E.I. and Haaf, T. Nuclear foci of mammalian recombination proteins are located at single-stranded DNA regions formed after DNA damage. *Proc Natl Acad Sci USA* 96, 1921-6 (1999).

363. Lisby, M. and Rothstein, R. Localization of checkpoint and repair proteins in eukaryotes. *Biochimie* 87, 579-89 (2005).

364. Hediger, F., Neumann, F.R., Van Houwe, G., Dubrana, K. and Gasser, S.M. Live imaging of telomeres: yKu and Sir proteins define redundant telomere-anchoring pathways in yeast. *Curr Biol* 12, 2076-89 (2002).

365. Corredor, E., Lukaszewski, A.J., Pachon, P., Allen, D.C. and Naranjo, T. Terminal regions of wheat chromosomes select their pairing partners in meiosis. *Genetics* 177, 699-706 (2007).

366. Moralli, D., Chan, D.Y., Jefferson, A., Volpi, E.V. and Monaco, Z.L. HAC stability in murine cells is influenced by nuclear localization and chromatin organization. *BMC Cell Biol* 10, 18 (2009).

367. Taddei, A. et al. The functional importance of telomere clustering: global changes in gene expression result from SIR factor dispersion. *Genome Res* 19, 611-25 (2009).

368. Sakuno, T. and Watanabe, Y. Studies of meiosis disclose distinct roles of cohesion in the core centromere and pericentromeric regions. *Chromosome Res* 17, 239-49 (2009).

369. Verschure, P.J. Positioning the genome within the nucleus. *Biol Cell* 96, 569-77 (2004).

370. Mora, L., Sanchez, I., Garcia, M. and Ponsa, M. Chromosome territory positioning of conserved homologous chromosomes in different primate species. *Chromosoma* 115, 367-75 (2006).

371. Kumaran, R.I., Thakar, R. and Spector, D.L. Chromatin dynamics and gene positioning. *Cell* 132, 929-34 (2008).

372. Simonis, M. and de Laat, W. FISH-eyed and genome-wide views on the spatial organisation of gene expression. *Biochim Biophys Acta* 1783, 2052-60 (2008).

373. Sengupta, K. et al. Position of human chromosomes is conserved in mouse nuclei indicating a species-independent mechanism for maintaining genome organization. *Chromosoma* 117, 499-509 (2008).

374. Ferrai, C., de Castro, I.J., Lavitas, L., Chotalia, M. and Pombo, A. Gene positioning. *Cold Spring Harb Perspect Biol* 2, a000588 (2010).

375. Parada, L.A., McQueen, P.G. and Misteli, T. Tissue-specific spatial organization of genomes. *Genome Biol* 5, R44 (2004).

376. Rosa, A. and Everaers, R. Structure and dynamics of interphase chromosomes. *PLoS Comput Biol* 4, e1000153 (2008).

377. Britten, R.J. and Kohne, D.E. Repeated Segments of DNA. *Sci Am* 222, 24-31 (1970).

378. Schmidt, T. LINEs, SINEs and repetitive DNA: non-LTR retrotransposons in plant genomes. *Plant Mol Biol* 40, 903-10 (1999).

379. Sharma, S. and Raina, S.N. Organization and evolution of highly repeated satellite DNA sequences in plant chromosomes. *Cytogenet Genome Res* 109, 15-26 (2005).

380. Richard, G.F., Kerrest, A. and Dujon, B. Comparative genomics and molecular dynamics of DNA repeats in eukaryotes. *Microbiol Mol Biol Rev* 72, 686-727 (2008).

381. Feschotte, C., Keswani, U., Ranganathan, N., Guibotsy, M.L. and Levine, D. Exploring repetitive DNA landscapes using REPCLASS, a tool that automates the classification of transposable elements in eukaryotic genomes. *Genome Biol Evol* 205-20 (2009).

382. Lander, E.S. et al. Initial sequencing and analysis of the human genome. *Nature* 409, 860-921 (2001).

383. Saunders, N.J. et al. Repeat-associated phase variable genes in the complete genome sequence of Neisseria meningitidis strain MC58. *Mol Microbiol* 37, 207-15 (2000).

384. Ugarkovic, D. Functional elements residing within satellite DNAs. *EMBO Rep* 6, 1035-9 (2005).

385. Sternberg, R. von and Shapiro, J.A. How repeated retroelements format genome function. *Cytogenet Genome Res* 110, 108-16 (2005).

386. Nishihara, H. and Okada, N. Retroposons: genetic footprints on the evolutionary paths of life. *Methods Mol Biol* 422, 201-25 (2008).

387. Bosco, G., Campbell, P., Leiva-Neto, J.T. and Markow, T.A. Analysis of Drosophila species genome size and satellite DNA content reveals significant differences among strains as well as between species. *Genetics* 177, 1277-90 (2007).

388. Kuhn, G.C., Sene, F.M., Moreira-Filho, O., Schwarzacher, T. and Heslop-Harrison, J.S. Sequence analysis, chromosomal distribution and long-range organization show that rapid turnover of new and old pBuM satellite DNA repeats leads to different patterns of variation in seven species of the Drosophila buzzatii cluster. *Chromosome Res* 16, 307-24 (2008).

389. Ferree, P.M. and Barbash, D.A. Species-Specific Heterochromatin Prevents Mitotic Chromosome Segregation to Cause Hybrid Lethality in Drosophila. *PLoS Biol* 7, e1000234 (2009).

390. Moller-Krull, M. et al. Retroposed elements and their flanking regions resolve the evolutionary history of xenarthran mammals (armadillos, anteaters, and sloths). *Mol Biol Evol* 24, 2573-82 (2007).

391. Churakov, G. et al. Rodent evolution: back to the root. *Mol Biol Evol* 27, 1315-26 (2010).

392. Berg, J.M., Tymoczko, J.L. and Stryer, L. *Biochemistry*, (W. H. Freeman and Co, New York, 2002).

393. Griffith, F. The Significance of Pneumococcal Types. *J Hyg (Lond)* 27, 113-59 (1928).

394. Avery, O.T., MacLeod, C.M., McCarty, M. Studies on the chemical nature of the substance inducing transformation of Pneumococcal types: Induction of transformation by a desoxyribonucleic acid fraction isolated prom Pneumococcus Type III. *J Exp Med* 79, 137-158 (1944).

395. Dubnau, D. DNA uptake in bacteria. *Annu Rev Microbiol* 53, 217-44 (1999).

396. Chen, I. and Dubnau, D. DNA uptake during bacterial transformation. *Nat Rev Microbiol* 2, 241-249 (2004).

397. Hayes, W. *The Genetics of Bacteria and Their Viruses (2nd ed.)*, (Blackwell, London, 1968).

398. Prangishvili, D. et al. Conjugation in archaea: frequent occurrence of conjugative plasmids in Sulfolobus. *Plasmid* 40, 190-202 (1998).

399. Juhas, M., Crook, D.W. and Hood, D.W. Type IV secretion systems: tools of bacterial horizontal gene transfer and virulence. *Cell Microbiol* 10, 2377-86 (2008).

400. Frank, A.C., Alsmark, C.M., Thollesson, M. and Andersson, S.G. Functional divergence and horizontal transfer of type IV secretion systems. *Mol Biol Evol* 22, 1325-36 (2005).

401. Averhoff, B. and Friedrich, A. Type IV pili-related natural transformation systems: DNA transport in mesophilic and thermophilic bacteria. *Arch Microbiol* 180, 385-93 (2003).

402. Cascales, E. and Christie, P.J. The versatile bacterial type IV secretion systems. *Nat Rev Microbiol* 1, 137-49 (2003).

403. Alvarez-Martinez, C.E. and Christie, P.J. Biological diversity of prokaryotic type IV secretion systems. *Microbiol Mol Biol Rev* 73, 775-808 (2009).

404. Heinemann, J.A. and Sprague, G.F., Jr. Bacterial conjugative plasmids mobilize DNA transfer between bacteria and yeast. *Nature* 340, 205-9 (1989).

405. Zupan, J., Muth, T.R., Draper, O. and Zambryski, P. The transfer of DNA from agrobacterium tumefaciens into plants: a feast of fundamental insights. *Plant J* 23, 11-28 (2000).

406. Yoon, Y.G. and Koob, M.D. Transformation of isolated mammalian mitochondria by bacterial conjugation. *Nucleic Acids Res* 33, e139 (2005).

407. Waters, V.L. Conjugation between bacterial and mammalian cells. *Nat Genet* 29, 375-6 (2001).

408. Zupan, J.R. and Zambryski, P. Transfer of T-DNA from Agrobacterium to the plant cell. *Plant Physiol* 107, 1041-7 (1995).

409. Gelvin, S.B. Agrobacterium-mediated plant transformation: the biology behind the "gene-jockeying" tool. *Microbiol Mol Biol Rev* 67, 16-37 (2003).

410. Broothaerts, W. et al. Gene transfer to plants by diverse species of bacteria. *Nature* 433, 629-33 (2005).

411. Tzfira, T. and Citovsky, V. Agrobacterium-mediated genetic transformation of plants: biology and biotechnology. *Curr Opin Biotechnol* 17, 147-54 (2006).

412. Haber, J.E., Ira, G., Malkova, A. and Sugawara, N. Repairing a double-strand chromosome break by homologous recombination: revisiting Robin Holliday's model. *Philos Trans R Soc Lond B Biol Sci* 359, 79-86 (2004).

413. Cavalier-Smith, T. Origins of the machinery of recombination and sex. *Heredity* 88, 125-41 (2002).

414. Page, S.L. and Hawley, R.S. The genetics and molecular biology of the synaptonemal complex. *Annu Rev Cell Dev Biol* 20, 525-58 (2004).

415. Zetka, M. Homologue pairing, recombination and segregation in Caenorhabditis elegans. *Genome Dyn* 5, 43-55 (2009).

416. Lao, J.P. and Hunter, N. Trying to Avoid Your Sister. *PLoS Biol* 8, e1000519 (2010).

417. Udall, J.A., Quijada, P.A. and Osborn, T.C. Detection of chromosomal rearrangements derived from homologous recombination in four mapping populations of Brassica napus L. *Genetics* 169, 967-79 (2005).

418. Lupski, J.R. and Stankiewicz, P. Genomic disorders: molecular mechanisms for rearrangements and conveyed phenotypes. *PLoS Genet* 1, e49 (2005).

419. Mieczkowski, P.A., Lemoine, F.J. and Petes, T.D. Recombination between retrotransposons as a source of chromosome rearrangements in the yeast Saccharomyces cerevisiae. *DNA Repair (Amst)* 5, 1010-20 (2006).

420. Gu, W., Zhang, F. and Lupski, J.R. Mechanisms for human genomic rearrangements. *Pathogenetics* 1, 4 (2008).

421. Argueso, J.L. et al. Double-strand breaks associated with repetitive DNA can reshape the genome. *Proc Natl Acad Sci USA* 105, 11845-50 (2008).

422. Lewis, L.K. and Resnick, M.A. Tying up loose ends: nonhomologous end-joining in Saccharomyces cerevisiae. *Mutat Res* 451, 71-89 (2000).

423. Pastwa, E. and Blasiak, J. Non-homologous DNA end joining. *Acta Biochim Pol* 50, 891-908 (2003).

424. Weterings, E. and Chen, D.J. The endless tale of non-homologous end-joining. *Cell Res* 18, 114-24 (2008).

425. Stephanou, N.C. et al. Mycobacterial nonhomologous end joining mediates mutagenic repair of chromosomal double-strand DNA breaks. *J Bacteriol* 189, 5237-46 (2007).

426. Zhuang, J., Jiang, G., Willers, H. and Xia, F. Exonuclease function of human Mre11 promotes deletional nonhomologous end joining. *J Biol Chem* 284, 30565-73 (2009).

427. Yu, X. and Gabriel, A. Reciprocal translocations in Saccharomyces cerevisiae formed by nonhomologous end joining. *Genetics* 166, 741-51 (2004).

428. Haber, J.E. Transpositions and translocations induced by site-specific double-strand breaks in budding yeast. *DNA Repair (Amst)* 5, 998-1009 (2006).

429. Lisby, M. and Rothstein, R. Choreography of recombination proteins during the DNA damage response. *DNA Repair (Amst)* 8, 1068-76 (2009).

430. Gopaul, D.N. and Duyne, G.D. Structure and mechanism in site-specific recombination. *Curr Opin Struct Biol* 9, 14-20 (1999).

431. Grindley, N.D., Whiteson, K.L. and Rice, P.A. Mechanisms of site-specific recombination. *Annu Rev Biochem* 75, 567-605 (2006).

432. Van Duyne, G.D. A structural view of cre-loxp site-specific recombination. *Annu Rev Biophys Biomol Struct* 30, 87-104 (2001).

433. Ghosh, K. and Van Duyne, G.D. Cre-loxP biochemistry. *Methods Cell Biol* 28, 374-83 (2002).

434. Chen, Y. and Rice, P.A. New insight into site-specific recombination from Flp recombinase-DNA structures. *Annu Rev Biophys Biomol Struct* 32, 135-59 (2003).

435. Segall, A.M. and Craig, N.L. New wrinkles and folds in site-specific recombination. *Mol Cell* 19, 433-5 (2005).

436. Smith, M.C. and Thorpe, H.M. Diversity in the serine recombinases. *Mol Microbiol* 44, 299-307 (2002).

437. Weinacht, K.G. et al. Tyrosine site-specific recombinases mediate DNA inversions affecting the expression of outer surface proteins of Bacteroides fragilis. *Mol Microbiol* 53, 1319-30 (2004).

438. Manson, J.M. and Gilmore, M.S. Pathogenicity island integrase cross-talk: a potential new tool for virulence modulation. *Mol Microbiol* 61, 555-9 (2006).

439. Juhas, M. et al. Genomic islands: tools of bacterial horizontal gene transfer and evolution. *FEMS Microbiol Rev* 33, 376-93 (2009).

440. Rolland, T., Neuveglise, C., Sacerdot, C. and Dujon, B. Insertion of horizontally transferred genes within conserved syntenic regions of yeast genomes. *PLoS One* 4, e6515 (2009).

441. McClintock, B. The origin and behavior of mutable loci in maize. *Proc Natl Acad Sci USA* 36, 344-55 (1950).

442. McClintock, B. Induction of Instability at Selected Loci in Maize. *Genetics* 38, 579-99 (1953).

443. Shapiro, J.A. Letting Escherichia coli teach me about genome engineering. *Genetics* 183, 1205-14 (2009).

444. Siguier, P., Filee, J. and Chandler, M. Insertion sequences in prokaryotic genomes. *Curr Opin Microbiol* 9, 526-31 (2006).

445. Bessereau, J.L. Transposons in C. elegans. *WormBook* 1-13 (2006).

446. Filee, J., Siguier, P. and Chandler, M. Insertion sequence diversity in archaea. *Microbiol Mol Biol Rev* 71, 121-57 (2007).

447. Parks, A.R. and Peters, J.E. Tn7 elements: engendering diversity from chromosomes to episomes. *Plasmid* 61, 1-14 (2009).

448. Peters, J.E. and Craig, N.L. Tn7 recognizes transposition target structures associated with DNA replication using the DNA-binding protein TnsE. *Genes Dev* 15, 737-47 (2001).

449. Posey, J.E., Pytlos, M.J., Sinden, R.R. and Roth, D.B. Target DNA structure plays a critical role in RAG transposition. *PLoS Biol* 4, e350 (2006).

450. Tobes, R. and Pareja, E. Bacterial repetitive extragenic palindromic sequences are DNA targets for Insertion Sequence elements. *BMC Genomics* 7, 62 (2006).

451. Mizuuchi, K. Mechanism of transposition of bacteriophage Mu: polarity of the strand transfer reaction at the initiation of transposition. *Cell* 39, 395-404 (1984).

452. Haniford, D.B. and Chaconas, G. Mechanistic aspects of DNA transposition. *Curr Opin Genet Dev* 2, 698-704 (1992).

453. Chaconas, G. Studies on a "jumping gene machine": higher-order nucleoprotein complexes in Mu DNA transposition. *Biochem Cell Biol* 77, 487-91 (1999).

454. Hickman, A.B., Chandler, M. and Dyda, F. Integrating prokaryotes and eukaryotes: DNA transposases in light of structure. *Crit Rev Biochem Mol Biol* 45, 50-69 (2010).

455. Duval-Valentin, G., Marty-Cointin, B. and Chandler, M. Requirement of IS911 replication before integration defines a new bacterial transposition pathway. *Embo J* 23, 3897-906 (2004).

456. Shapiro, J.A. Molecular model for the transposition and replication of bacteriophage Mu and other transposable elements. *Proc Natl Acad Sci USA* 76, 1933-7 (1979).

457. Roberts, A.P. et al. Revised nomenclature for transposable genetic elements. *Plasmid* 60, 167-73 (2008).

458. Chandler, M., Roulet, E., Silver, L., Boy de la Tour, E. and Caro, L. Tn10 mediated integration of the plasmid R100.1 into the bacterial chromosome: inverse transposition. *Mol Gen Genet* 173, 23-30 (1979).

459. Muster, C.J., MacHattie, L.A. and Shapiro, J.A. pλCM system: observations on the roles of transposable elements in formation and breakdown of plasmids derived from bacteriophage λ replicons. *J Bacteriol* 153, 976-90 (1983).

460. Nag, D.K. et al. IS50-mediated inverse transposition: specificity and precision. *Gene* 34, 17-26 (1985).

461. Engels, W.R. Gene Duplication. *Science* 214, 786-787 (1981).

462. Engels, W.R. and Preston, C.R. Formation of chromosome rearrangements by P factors in Drosophila. *Genetics* 107, 657-78 (1984).

463. Preston, C.R., Sved, J.A. and Engels, W.R. Flanking duplications and deletions associated with P-induced male recombination in Drosophila. *Genetics* 144, 1623-38 (1996).

464. Gray, Y.H. It takes two transposons to tango: transposable-element-mediated chromosomal rearrangements. *Trends Genet* 16, 461-8 (2000).

465. Laski, F.A., Rio, D.C. and Rubin, G.M. Tissue specificity of Drosophila P element transposition is regulated at the level of mRNA splicing. *Cell* 44, 7-19 (1986).

466. Siebel, C.W. and Rio, D.C. Regulated splicing of the Drosophila P transposable element third intron in vitro: somatic repression. *Science* 248, 1200-8 (1990).

467. Rho, M., Choi, J.H., Kim, S., Lynch, M. and Tang, H. De novo identification of LTR retrotransposons in eukaryotic genomes. *BMC Genomics* 8, 90 (2007).

468. Llorens, C., Munoz-Pomer, A., Bernad, L., Botella, H. and Moya, A. Network dynamics of eukaryotic LTR retroelements beyond phylogenetic trees. *Biol Direct* 4, 41 (2009).

469. Coffin, J.M., Hughes, S.H. and Varmus, H.E. *Retroviruses* (Cold Spring Harbor Laboratory Press, Cold Spring Harbor, New York, 1997).

470. Varmus, H. Reverse Transcription. *Sci Am* 257, 56-64 (1987).

471. Temin, H. The participation of DNA in rous sarcoma virus production. *Virology* 23, 486-494 (1964).

472. Lwoff, A. Lysogeny. *Bacteriol Rev* 17, 269-337 (1953).

473. Lwoff, A. The Life Cycle of a Virus. *Sci Am* 190, 34-37 (1954).

474. Lwoff, A. Interaction among virus, cell, and organism. *Science* 152, 1216-20 (1966).

475. Cheng, Z. and Menees, T.M. RNA Branching and Debranching in the Yeast Retrovirus-like Element Ty1. *Science* 303, 240-243 (2004).

476. Eickbush, T.H. and Jamburuthugoda, V.K. The diversity of retrotransposons and the properties of their reverse transcriptases. *Virus Res* 134, 221-34 (2008).

477. Basu, V.P. et al. Strand transfer events during HIV-1 reverse transcription. *Virus Res* 134, 19-38 (2008).

478. Engelman, A., Mizuuchi, K. and Craigie, R. HIV-1 DNA integration: mechanism of viral DNA cleavage and DNA strand transfer. *Cell* 67, 1211-21 (1991).

479. Polard, P. and Chandler, M. Bacterial transposases and retroviral integrases. *Mol Microbiol* 15, 13-23 (1995).

480. Li, M., Mizuuchi, M., Burke, T.R., Jr. and Craigie, R. Retroviral DNA integration: reaction pathway and critical intermediates. *Embo J* 25, 1295-304 (2006).

481. Goodwin, T.J. and Poulter, R.T. The DIRS1 group of retrotransposons. *Mol Biol Evol* 18, 2067-82 (2001).

482. Poulter, R.T. and Goodwin, T.J. DIRS-1 and the other tyrosine recombinase retrotransposons. *Cytogenet Genome Res* 110, 575-88 (2005).

483. Rous, P. A transmissable avian neoplasm. *J Exp Med* 12, 696-705 (1910).

484. Van Epps, H.L. Peyton Rous, father of the tumor virus. *J Exp Med* 201, 320 (2005).

485. Maeda, N., Fan, H. and Yoshikai, Y. Oncogenesis by retroviruses: old and new paradigms. *Rev Med Virol* 18, 387-405 (2008).

486. Swain, A. and Coffin, J.M. Mechanism of transduction by retroviruses. *Science* 255, 841-5 (1992).

487. Tsichlis, P.N., Strauss, P.G. and Hu, L.F. A common region for proviral DNA integration in MoMuLV-induced rat thymic lymphomas. *Nature* 302, 445-9 (1983).

488. Tsichlis, P.N. Oncogenesis by Moloney murine leukemia virus. *Anticancer Res* 7, 171-80 (1987).

489. Uren, A.G., Kool, J., Berns, A. and van Lohuizen, M. Retroviral insertional mutagenesis: past, present and future. *Oncogene* 24, 7656-72 (2005).

490. Mellentin-Michelotti, J., John, S., Pennie, W.D., Williams, T. and Hager, G.L. The 5' enhancer of the mouse mammary tumor virus long terminal repeat contains a functional AP-2 element. *J Biol Chem* 269, 31983-90 (1994).

491. Errede, B. et al. Mating signals control expression of mutations resulting from insertion of a transposable repetitive element adjacent to diverse yeast genes. *Cell* 22, 427-36 (1980).

492. Errede, B. et al. Studies on transposable elements in yeast. I. ROAM mutations causing increased expression of yeast genes: their activation by signals directed toward conjugation functions and their formation by insertion of Ty1 repetitive elements. II. deletions, duplications, and transpositions of the COR segment that encompasses the structural gene of yeast iso-1-cytochrome c. *Cold Spring Harb Symp Quant Biol* 45 Pt 2, 593-607 (1981).

493. Modolell, J., Bender, W. and Meselson, M. Drosophila melanogaster mutations suppressible by the suppressor of Hairy-wing are insertions of a 7.3-kilobase mobile element. *Proc Natl Acad Sci USA* 80, 1678-82 (1983).

494. Peifer, M. and Bender, W. Sequences of the gypsy transposon of Drosophila necessary for its effects on adjacent genes. *Proc Natl Acad Sci USA* 85, 9650-4 (1988).

495. Harrison, D.A., Geyer, P.K., Spana, C. and Corces, V.G. The gypsy retrotransposon of Drosophila melanogaster: mechanisms of mutagenesis and interaction with the suppressor of Hairy-wing locus. *Dev Genet* 10, 239-48 (1989).

496. Corces, V.G. and Geyer, P.K. Interactions of retrotransposons with the host genome: the case of the gypsy element of Drosophila. *Trends Genet* 7, 86-90 (1991).

497. Ilyin, Y.V., Lyubomirskaya, N.V. and Kim, A.I. Retrotransposon Gypsy and genetic instability in Drosophila (review). *Genetica* 85, 13-22 (1991).

498. Pelisson, A., Mejlumian, L., Robert, V., Terzian, C. and Bucheton, A. Drosophila germline invasion by the endogenous retrovirus gypsy: involvement of the viral env gene. *Insect Biochem Mol Biol* 32, 1249-56 (2002).

499. Gdula, D.A., Gerasimova, T.I. and Corces, V.G. Genetic and molecular analysis of the gypsy chromatin insulator of Drosophila. *Proc Natl Acad Sci USA* 93, 9378-83 (1996).

500. Cai, H.N. and Levine, M. The gypsy insulator can function as a promoter-specific silencer in the Drosophila embryo. *Embo J* 16, 1732-41 (1997).

501. Chen, S. and Corces, V.G. The gypsy insulator of Drosophila affects chromatin structure in a directional manner. *Genetics* 159, 1649-58 (2001).

502. Byrd, K. and Corces, V.G. Visualization of chromatin domains created by the gypsy insulator of Drosophila. *J Cell Biol* 162, 565-74 (2003).

503. Fawcett, D.H., Lister, C.K., Kellett, E. and Finnegan, D.J. Transposable elements controlling I-R hybrid dysgenesis in D. melanogaster are similar to mammalian LINEs. *Cell* 47, 1007-15 (1986).

504. Bucheton, A. I transposable elements and I-R hybrid dysgenesis in Drosophila. *Trends Genet* 6, 16-21 (1990).

505. Busseau, I., Chaboissier, M.C., Pelisson, A. and Bucheton, A. I factors in Drosophila melanogaster: transposition under control. *Genetica* 93, 101-16 (1994).

506. Lyozin, G.T. et al. The structure and evolution of Penelope in the virilis species group of Drosophila: an ancient lineage of retroelements. *J Mol Evol* 52, 445-56 (2001).

507. Yang, H.P. and Barbash, D.A. Abundant and species-specific DINE-1 transposable elements in 12 Drosophila genomes. *Genome Biol* 9, R39 (2008).

508. Sezutsu, H., Nitasaka, E. and Yamazaki, T. Evolution of the LINE-like I element in the Drosophila melanogaster species subgroup. *Mol Gen Genet* 249, 168-78 (1995).

509. Brosius, J. RNAs from all categories generate retrosequences that may be exapted as novel genes or regulatory elements. *Gene* 238, 115-134 (1999).

510. Pavlicek, A., Gentles, A.J., Paces, J., Paces, V. and Jurka, J. Retroposition of processed pseudogenes: the impact of RNA stability and translational control. *Trends Genet* 22, 69-73 (2006).

511. Druker, R. and Whitelaw, E. Retrotransposon-derived elements in the mammalian genome: a potential source of disease. *J Inherit Metab Dis* 27, 319-30 (2004).

512. Gogvadze, E. and Buzdin, A. Retroelements and their impact on genome evolution and functioning. *Cell Mol Life Sci* 66, 3727-42 (2009).

513. Goodier, J.L., Ostertag, E.M. and Kazazian, H.H., Jr. Transduction of 3'-flanking sequences is common in L1 retrotransposition. *Hum Mol Genet* 9, 653-7 (2000).

514. Moran, J.V., DeBerardinis, R.J. and Kazazian, H.H., Jr. Exon shuffling by L1 retrotransposition. *Science* 283, 1530-4 (1999).

515. Szak, S.T., Pickeral, O.K., Landsman, D. and Boeke, J.D. Identifying related L1 retrotransposons by analyzing 3' transduced sequences. *Genome Biol* 4, R30 (2003).

516. Ejima, Y. and Yang, L. Trans mobilization of genomic DNA as a mechanism for retrotransposon-mediated exon shuffling. *Hum Mol Genet* 12, 1321-8 (2003).

517. Lovering, R., Harden, N. and Ashburner, M. The molecular structure of TE146 and its derivatives in Drosophila melanogaster. *Genetics* 128, 357-72 (1991).

518. Mills, R.E. et al. Recently mobilized transposons in the human and chimpanzee genomes. *Am J Hum Genet* 78, 671-9 (2006).

519. Bantysh, O.B. and Buzdin, A.A. Novel Family of Human Transposable Elements Formed Due to Fusion of the First Exon of Gene MAST2 with Retrotransposon SVA. *Biochemistry (Mosc)* 74, 1393-9 (2009).

520. Wang, H. et al. SVA elements: a hominid-specific retroposon family. *J Mol Biol* 354, 994-1007 (2005).

521. Ostertag, E.M., Goodier, J.L., Zhang, Y. and Kazazian, H.H., Jr. SVA elements are nonautonomous retrotransposons that cause disease in humans. *Am J Hum Genet* 73, 1444-51 (2003).

522. Hancks, D.C., Ewing, A.D., Chen, J.E., Tokunaga, K. and Kazazian, H.H., Jr. Exon-trapping mediated by the human retrotransposon SVA. *Genome Res* 19, 1983-91 (2009).

523. Damert, A. et al. 5'-Transducing SVA retrotransposon groups spread efficiently throughout the human genome. *Genome Res* 19, 1992-2008 (2009).

524. Xing, J. et al. Mobile elements create structural variation: analysis of a complete human genome. *Genome Res* 19, 1516-26 (2009).

525. Mills, R.E., Bennett, E.A., Iskow, R.C. and Devine, S.E. Which transposable elements are active in the human genome? *Trends Genet* 23, 183-91 (2007).

526. Ferat, J.L. and Michel, F. Group II self splicing introns in bacteria. *Nature* 364, 358-361 (1993).

527. Dunny, G.M. and McKay, L.L. Group II introns and expression of conjugative transfer functions in lactic acid bacteria. *Antonie Van Leeuwenhoek* 76, 77-88 (1999).

528. Lambowitz, A.M. and Zimmerly, S. Mobile group II introns. *Annu Rev Genet* 38, 1-35 (2004).

529. Valles, Y., Halanych, K.M. and Boore, J.L. Group II introns break new boundaries: presence in a bilaterian's genome. *PLoS One* 3, e1488 (2008).

530. Mohr, G. and Lambowitz, A.M. Putative proteins related to group II intron reverse transcriptase/maturases are encoded by nuclear genes in higher plants. *Nucleic Acids Res* 31, 647-52 (2003).

531. Cech, T.R. RNA as an Enzyme. *Sci Am* 255, 64-75 (1986).

532. Roitzsch, M. and Pyle, A.M. The linear form of a group II intron catalyzes efficient autocatalytic reverse splicing, establishing a potential for mobility. *RNA* 15, 473-82 (2009).

533. Fedorova, O. and Zingler, N. Group II introns: structure, folding and splicing mechanism. *Biol Chem* 388, 665-78 (2007).

534. Lehmann, K. and Schmidt, U. Group II introns: structure and catalytic versatility of large natural ribozymes. *Crit Rev Biochem Mol Biol* 38, 249-303 (2003).

535. Morl, M. and Schmelzer, C. Integration of group II intron bI1 into a foreign RNA by reversal of the self-splicing reaction in vitro. *Cell* 60, 629-36 (1990).

536. Eickbush, T.H. Mobile introns: retrohoming by complete reverse splicing. *Curr Biol* 9, R11-4 (1999).

537. Eskes, R. et al. Multiple homing pathways used by yeast mitochondrial group II introns. *Mol Cell Biol* 20, 8432-46 (2000).

538. Zimmerly, S., Guo, H., Perlman, P.S. and Lambowitz, A.M. Group II intron mobility occurs by target DNA-primed reverse transcription. *Cell* 82, 545-54 (1995).

539. Mohr, S., Matsuura, M., Perlman, P.S. and Lambowitz, A.M. A DEAD-box protein alone promotes group II intron splicing and reverse splicing by acting as an RNA chaperone. *Proc Natl Acad Sci USA* 103, 3569-74 (2006).

540. Smith, D., Zhong, J., Matsuura, M., Lambowitz, A.M. and Belfort, M. Recruitment of host functions suggests a repair pathway for late steps in group II intron retrohoming. *Genes Dev* 19, 2477-87 (2005).

541. Watanabe, K. and Lambowitz, A.M. High-affinity binding site for a group II intron-encoded reverse transcriptase/maturase within a stem-loop structure in the intron RNA. *RNA* 10, 1433-43 (2004).

542. Cui, X., Matsuura, M., Wang, Q., Ma, H. and Lambowitz, A.M. A group II intron-encoded maturase functions preferentially in cis and requires both the reverse transcriptase and X domains to promote RNA splicing. *J Mol Biol* 340, 211-31 (2004).

543. Toor, N., Rajashankar, K., Keating, K.S. and Pyle, A.M. Structural basis for exon recognition by a group II intron. *Nat Struct Mol Biol* 15, 1221-2 (2008).

544. Dickson, L. et al. Retrotransposition of a yeast group II intron occurs by reverse splicing directly into ectopic DNA sites. *Proc Natl Acad Sci USA* 98, 13207-12 (2001).

545. Costa, M., Michel, F. and Toro, N. Potential for alternative intron-exon pairings in group II intron RmInt1 from Sinorhizobium meliloti and its relatives. *RNA* 12, 338-41 (2006).

546. Zhong, J. and Lambowitz, A.M. Group II intron mobility using nascent strands at DNA replication forks to prime reverse transcription. *Embo J* 22, 4555-65 (2003).

547. Robart, A.R., Seo, W. and Zimmerly, S. Insertion of group II intron retroelements after intrinsic transcriptional terminators. *Proc Natl Acad Sci USA* 104, 6620-5 (2007).

548. Karberg, M. et al. Group II introns as controllable gene targeting vectors for genetic manipulation of bacteria. *Nat Biotechnol* 19, 1162-7 (2001).

549. Zhong, J., Karberg, M. and Lambowitz, A.M. Targeted and random bacterial gene disruption using a group II intron (targetron) vector containing a retrotransposition-activated selectable marker. *Nucleic Acids Res* 31, 1656-64 (2003).

550. Jones, J.P., 3rd et al. Retargeting mobile group II introns to repair mutant genes. *Mol Ther* 11, 687-94 (2005).

551. Guo, H. et al. Group II introns designed to insert into therapeutically relevant DNA target sites in human cells. *Science* 289, 452-7 (2000).

552. Rawsthorne, H., Turner, K.N. and Mills, D.A. Multicopy integration of heterologous genes, using the lactococcal group II intron targeted to bacterial insertion sequences. *Appl Environ Microbiol* 72, 6088-93 (2006).

553. Mohr, G., Ghanem, E. and Lambowitz, A.M. Mechanisms used for genomic proliferation by thermophilic group II introns. *PLoS Biol* 8, e1000391 (2010).

554. Liu, X.Q. Protein-splicing intein: Genetic mobility, origin, and evolution. *Annu Rev Genet* 34, 61-76 (2000).

555. Gogarten, J.P., Senejani, A.G., Zhaxybayeva, O., Olendzenski, L. and Hilario, E. Inteins: structure, function, and evolution. *Annu Rev Microbiol* 56, 263-87 (2002).

556. Elleuche, S. and Poggeler, S. Inteins, valuable genetic elements in molecular biology and biotechnology. *Appl Microbiol Biotechnol* 87, 479-89 (2010).

557. Gimble, F.S. Invasion of a multitude of genetic niches by mobile endonuclease genes. *FEMS Microbiol Lett* 185, 99-107 (2000).

558. Gogarten, J.P. and Hilario, E. Inteins, introns, and homing endonucleases: recent revelations about the life cycle of parasitic genetic elements. *BMC Evol Biol* 6, 94 (2006).

559. Raghavan, R. and Minnick, M.F. Group I introns and inteins: disparate origins but convergent parasitic strategies. *J Bacteriol* 191, 6193-202 (2009).

560. Perler, F.B. InBase: the Intein Database. *Nucleic Acids Res* 30, 383-4 (2002).

561. Goodwin, T.J., Butler, M.I. and Poulter, R.T. Multiple, non-allelic, inteincoding sequences in eukaryotic RNA polymerase genes. *BMC Biol* 4, 38 (2006).

562. Liu, M. et al. Reverse transcriptase-mediated tropism switching in Bordetella bacteriophage. *Science* 295, 2091-4 (2002).

563. Doulatov, S. et al. Tropism switching in Bordetella bacteriophage defines a family of diversity-generating retroelements. *Nature* 431, 476-81 (2004).

564. Guo, H. et al. Diversity-generating retroelement homing regenerates target sequences for repeated rounds of codon rewriting and protein diversification. *Mol Cell* 31, 813-23 (2008).

565. Medhekar, B. and Miller, J.F. Diversity-generating retroelements. *Curr Opin Microbiol* 10, 388-95 (2007).

566. Wisniewski-Dye, F. and Vial, L. Phase and antigenic variation mediated by genome modifications. *Antonie Van Leeuwenhoek* 94, 493-515 (2008).

567. Silverman, M., Zieg, J., Hilmen, M. and Simon, M. Phase variation in Salmonella: genetic analysis of a recombinational switch. *Proc Natl Acad Sci USA* 76, 391-5 (1979).

568. Lysnyansky, I., Ron, Y. and Yogev, D. Juxtaposition of an active promoter to vsp genes via site-specific DNA inversions generates antigenic variation in Mycoplasma bovis. *J Bacteriol* 183, 5698-708 (2001).

569. Blomfield, I.C. The regulation of pap and type 1 fimbriation in Escherichia coli. *Adv Microb Physiol* 45, 1-49 (2001).

570. Emerson, J.E. et al. A novel genetic switch controls phase variable expression of CwpV, a Clostridium difficile cell wall protein. *Mol Microbiol* 74, 541-56 (2009).

571. Kutsukake, K., Nakashima, H., Tominaga, A. and Abo, T. Two DNA invertases contribute to flagellar phase variation in Salmonella enterica serovar Typhimurium strain LT2. *J Bacteriol* 188, 950-7 (2006).

572. Plasterk, R.H., Simon, M.I. and Barbour, A.G. Transposition of structural genes to an expression sequence on a linear plasmid causes antigenic variation in the bacterium Borrelia hermsii. *Nature* 318, 257-63 (1985).

573. Stern, A., Brown, M., Nickel, P. and Meyer, T.F. Opacity genes in Neisseria gonorrhoeae: control of phase and antigenic variation. *Cell* 47, 61-71 (1986).

574. Stern, A. and Meyer, T.F. Common mechanism controlling phase and antigenic variation in pathogenic Neisseriae. *Mol Microbiol* 1, 5-12 (1987).

575. Borst, P. and Greaves, D.R. Programmed gene rearrangements altering gene expression. *Science* 235, 658-67 (1987).

576. Barbour, A.G., Dai, Q., Restrepo, B.I., Stoenner, H.G. and Frank, S.A. Pathogen escape from host immunity by a genome program for antigenic variation. *Proc Natl Acad Sci USA* 103, 18290-5 (2006).

577. Komano, T. Shufflons: multiple inversion systems and integrons. *Annu Rev Genet* 33, 171-91 (1999).

578. Gyohda, A., Furuya, N., Ishiwa, A., Zhu, S. and Komano, T. Structure and function of the shufflon in plasmid R64. *Adv Biophys* 38, 183-213 (2004).

579. Tam, C.K., Hackett, J. and Morris, C. Rate of inversion of the Salmonella enterica shufflon regulates expression of invertible DNA. *Infect Immun* 73, 5568-77 (2005).

580. Cerdeno-Tarraga, A.M. et al. Extensive DNA inversions in the B. fragilis genome control variable gene expression. *Science* 307, 1463-5 (2005).

581. Campbell, A.M. How Viruses Insert their DNA into the DNA of the Host Cell. *Sci Am* 235, 102-113 (1976).

582. Nash, H.A. Integration and excision of bacteriophage lambda: the mechanism of conservation site specific recombination. *Annu Rev Genet* 15, 143-67 (1981).

583. Landy, A. Dynamic, structural, and regulatory aspects of lambda site-specific recombination. *Annu Rev Biochem* 58, 913-49 (1989).

584. Van Duyne, G.D. Lambda integrase: armed for recombination. *Curr Biol* 15, R658-60 (2005).

585. Radman-Livaja, M., Biswas, T., Ellenberger, T., Landy, A. and Aihara, H. DNA arms do the legwork to ensure the directionality of lambda site-specific recombination. *Curr Opin Struct Biol* 16, 42-50 (2006).

586. Barre, F.X. et al. Circles: the replication-recombination-chromosome segregation connection. *Proc Natl Acad Sci USA* 98, 8189-95 (2001).

587. Piggot, P.J. and Hilbert, D.W. Sporulation of Bacillus subtilis. *Curr Opin Microbiol* 7, 579-86 (2004).

588. Zhang, C.C., Laurent, S., Sakr, S., Peng, L. and Bedu, S. Heterocyst differentiation and pattern formation in cyanobacteria: a chorus of signals. *Mol Microbiol* 59, 367-75 (2006).

589. Stragier, P., Kunkel, B., Kroos, L. and Losick, R. Chromosomal rearrangement generating a composite gene for a developmental transcription factor. *Science* 243, 507-12 (1989).

590. Carrasco, C.D. and Golden, J.W. Two heterocyst-specific DNA rearrangements of nif operons in Anabaena cylindrica and Nostoc sp. strain Mac. *Microbiology* 141 (Pt 10), 2479-87 (1995).

591. Hallet, B. and Sherratt, D.J. Transposition and site-specific recombination: adapting DNA cut-and-paste mechanisms to a variety of genetic rearrangements. *FEMS Microbiol Rev* 21, 157-78 (1997).

592. Haber, J.E. Mating-type gene switching in Saccharomyces cerevisiae. *Annu Rev Genet* 32, 561-99 (1998).

593. Egel, R. Fission yeast mating-type switching: programmed damage and repair. *DNA Repair (Amst)* 4, 525-36 (2005).

594. Klar, A.J. Lessons learned from studies of fission yeast mating-type switching and silencing. *Annu Rev Genet* 41, 213-36 (2007).

595. Nasmyth, K. Regulating the HO endonuclease in yeast. *Curr Opin Genet Dev* 3, 286-94 (1993).

596. Haber, J.E. In vivo biochemistry: physical monitoring of recombination induced by site-specific endonucleases. *Bioessays* 17, 609-20 (1995).

597. Barsoum, E., Martinez, P. and Åström, S.U. α3, a transposable element that promotes host sexual reproduction. *Genes Dev* 24, 33-44 (2010).

598. Rusche, L.N. and Rine, J. Switching the mechanism of mating type switching: a domesticated transposase supplants a domesticated homing endonuclease. *Genes Dev* 24, 10-4 (2010).

599. Prescott, D.M. Genome gymnastics: unique modes of DNA evolution and processing in ciliates. *Nat Rev Genet* 1, 191-8 (2000).

600. Juranek, S.A. and Lipps, H.J. New insights into the macronuclear development in ciliates. *Int Rev Cytol* 262, 219-51 (2007).

601. Betermier, M. Large-scale genome remodelling by the developmentally programmed elimination of germ line sequences in the ciliate Paramecium. *Res Microbiol* 155, 399-408 (2004).

602. Chalker, D.L. Dynamic nuclear reorganization during genome remodeling of Tetrahymena. *Biochim Biophys Acta* 1783, 2130-6 (2008).

603. Jonsson, F., Postberg, J. and Lipps, H.J. The unusual way to make a genetically active nucleus. *DNA Cell Biol* 28, 71-8 (2009).

604. Mochizuki, K. and Gorovsky, M.A. Small RNAs in genome rearrangement in Tetrahymena. *Curr Opin Genet Dev* 14, 181-7 (2004).

605. Nowacki, M. et al. RNA-mediated epigenetic programming of a genome-rearrangement pathway. *Nature* 451, 153-8 (2008).

606. Meyer, E. and Duharcourt, S. Epigenetic regulation of programmed genomic rearrangements in Paramecium aurelia. *J Eukaryot Microbiol* 43, 453-61 (1996).

607. Garnier, O., Serrano, V., Duharcourt, S. and Meyer, E. RNA-mediated programming of developmental genome rearrangements in Paramecium tetraurelia. *Mol Cell Biol* 24, 7370-9 (2004).

608. Wong, L.C. and Landweber, L.F. Evolution of programmed DNA rearrangements in a scrambled gene. *Mol Biol Evol* 23, 756-63 (2006).

609. Mollenbeck, M. et al. The pathway to detangle a scrambled gene. *PLoS One* 3, e2330 (2008).

610. Duharcourt, S., Lepere, G. and Meyer, E. Developmental genome rearrangements in ciliates: a natural genomic subtraction mediated by non-coding transcripts. *Trends Genet* 25, 344-50 (2009).

611. Greider, C.W. and Blackburn, E.H. Identification of a specific telomere terminal transferase activity in Tetrahymena extracts. *Cell* 43, 405-13 (1985).

612. Greider, C.W. and Blackburn, E.H. A telomeric sequence in the RNA of Tetrahymena telomerase required for telomere repeat synthesis. *Nature* 337, 331-7 (1989).

613. Kano, H. et al. L1 retrotransposition occurs mainly in embryogenesis and creates somatic mosaicism. *Genes Dev* 23, 1303-12 (2009).

614. Muotri, A.R. and Gage, F.H. Generation of neuronal variability and complexity. *Nature* 441, 1087-93 (2006).

615. Muotri, A.R. et al. Somatic mosaicism in neuronal precursor cells mediated by L1 retrotransposition. *Nature* 435, 903-10 (2005).

616. Kuwabara, T. et al. Wnt-mediated activation of NeuroD1 and retro-elements during adult neurogenesis. *Nat Neurosci* 12, 1097-105 (2009).

617. Muotri, A.R., Zhao, C., Marchetto, M.C. and Gage, F.H. Environmental influ-ence on L1 retrotransposons in the adult hippocampus. *Hippocampus* 19, 1002-7 (2009).

618. Muotri, A.R., Marchetto, M.C., Coufal, N.G. and Gage, F.H. The necessary junk: new functions for transposable elements. *Hum Mol Genet* 16 Spec No. 2, R159-67 (2007).

619. Muotri, A.R. et al. L1 retrotransposition in neurons is modulated by MeCP2. *Nature* 468, 443-446 (2010).

620. Coufal, N.G. et al. L1 retrotransposition in human neural progenitor cells. *Nature* 460, 1127-31 (2009).

621. Singer, T., McConnell, M.J., Marchetto, M.C., Coufal, N.G. and Gage, F.H. LINE-1 retrotransposons: mediators of somatic variation in neuronal genomes? *Trends Neurosci* 33, 345-54 (2010).

622. Samaco, R.C., Nagarajan, R.P., Braunschweig, D. and LaSalle, J.M. Multiple pathways regulate MeCP2 expression in normal brain development and exhibit defects in autism-spectrum disorders. *Hum Mol Genet* 13, 629-39 (2004).

623. Peddada, S., Yasui, D.H. and LaSalle, J.M. Inhibitors of differentiation (ID1, ID2, ID3 and ID4) genes are neuronal targets of MeCP2 that are elevated in Rett syndrome. *Hum Mol Genet* 15, 2003-14 (2006).

624. Vanderhaeghen, P. Wnts blow on NeuroD1 to promote adult neuron produc-tion and diversity. *Nat Neurosci* 12, 1079-81 (2009).

625. Sarachana, T., Zhou, R., Chen, G., Manji, H.K. and Hu, V.W. Investigation of post-transcriptional gene regulatory networks associated with autism spectrum disorders by microRNA expression profiling of lymphoblastoid cell lines. *Genome Med* 2, 23 (2010).

626. St Laurent, G., 3rd, Hammell, N. and McCaffrey, T.A. A LINE-1 component to human aging: do LINE elements exact a longevity cost for evolutionary advantage? *Mech Ageing Dev* 131, 299-305 (2010).

627. Iskow, R.C. et al. Natural mutagenesis of human genomes by endogenous retrotransposons. *Cell* 141, 1253-61 (2010).

628. Burnet, M. How Antibodies are Made. *Sci Am* 191, 74-78 (1954).

629. Burnet, M. The Mechanism of Immunity. *Sci Am* 204, 58-67 (1961).

630. Ada, G.L. and Nossal, G. The clonal-selection theory. *Sci Am* 257, 62-9 (1987).

631. Tonegawa, S. The molecules of the immune system. *Sci Am* 253, 122-31 (1985).

632. Edelman, G.M. The structure and function of antibodies. *Sci Am* 223, 34-42 (1970).

633. Leder, P. The genetics of antibody diversity. *Sci Am* 246, 102-15 (1982).

634. Nossal, G.J.V. Life, Death and the Immune System. *Sci Am* 269, 52-62 (1993).

635. Clatworthy, A.E., Valencia, M.A., Haber, J.E. and Oettinger, M.A. V(D)J recombination and RAG-mediated transposition in yeast. *Mol Cell* 12, 489-99 (2003).

636. Lamrani, S. et al. Starvation-induced Mucts62-mediated coding sequence fusion: a role for ClpXP, Lon, RpoS and Crp. *Mol Microbiol* 32, 327-43 (1999).

637. Ilves, H., Horak, R., Teras, R. and Kivisaar, M. IHF is the limiting host factor in transposition of Pseudomonas putida transposon Tn4652 in stationary phase. *Mol Microbiol* 51, 1773-85 (2004).

638. Horak, R., Ilves, H., Pruunsild, P., Kuljus, M. and Kivisaar, M. The ColR-ColS two-component signal transduction system is involved in regulation of Tn4652 transposition in Pseudomonas putida under starvation conditions. *Mol Microbiol* 54, 795-807 (2004).

639. Kinsey, P.T. and Sandmeyer, S.B. Ty3 transposes in mating populations of yeast: a novel transposition assay for Ty3. *Genetics* 139, 81-94 (1995).

640. Ke, N., Irwin, P.A. and Voytas, D.F. The pheromone response pathway activates transcription of Ty5 retrotransposons located within silent chromatin of Saccharomyces cerevisiae. *Embo J* 16, 6272-80 (1997).

641. Sehgal, A., Lee, C.Y. and Espenshade, P.J. SREBP controls oxygen-dependent mobilization of retrotransposons in fission yeast. *PLoS Genet* 3, e131 (2007).

642. Shapiro, J.A. Observations on the formation of clones containing araB-lacZ cistron fusions. *Mol Gen Genet* 194, 79-90 (1984).

643. Stebbins, J., G.L. Cataclysmic Evolution. *Sci Am* 184, 54 -59 (1951).

644. McClintock, B. The significance of responses of the genome to challenge. *Science* 226, 792-801 (1984).

645. Feldman, M. and Levy, A.A. Allopolyploidy—a shaping force in the evolution of wheat genomes. *Cytogenet Genome Res* 109, 250-8 (2005).

646. Kidwell, M.G. Hybrid dysgenesis in Drosophila melanogaster: nature and inheritance of P element regulation. *Genetics* 111, 337-50 (1985).

647. Kidwell, M.G., Kimura, K. and Black, D.M. Evolution of hybrid dysgenesis potential following P element contamination in Drosophila melanogaster. *Genetics* 119, 815-28 (1988).

648. Bregliano, J. and Kidwell, M. Hybrid dysgenesis. In *Mobile Genetic Elements* (ed. Shapiro, J.) 363-410. (Academic Press, New York, 1983).

649. Woodruff, R.C. and Thomson, J.N. The fundamental theorem of neutral evolution: rates of substitution and mutation should factor in premeiotic clusters. *Genetica* 125, 333-9 (2005).

650. Kocur, G.J., Drier, E.A. and Simmons, M.J. Sterility and hypermutability in the P-M system of hybrid dysgenesis in Drosophila melanogaster. *Genetics* 114, 1147-63 (1986).

651. Thorp, M.W., Chapman, E.J. and Simmons, M.J. Cytotype regulation by telomeric P elements in Drosophila melanogaster: variation in regulatory strength and maternal effects. *Genet Res* 91, 327-36 (2009).

652. Jensen, P.A., Stuart, J.R., Goodpaster, M.P., Goodman, J.W. and Simmons, M.J. Cytotype regulation of P transposable elements in Drosophila melanogaster: repressor polypeptides or piRNAs? *Genetics* 179, 1785-93 (2008).

653. Simmons, M.J. et al. Cytotype regulation by telomeric P elements in Drosophila melanogaster: interactions with P elements from M' strains. *Genetics* 176, 1957-66 (2007).

654. Belinco, C. et al. Cytotype regulation in Drosophila melanogaster: synergism between telomeric and non-telomeric P elements. *Genet Res* 91, 383-94 (2009).

655. Brennecke, J. et al. Discrete small RNA-generating loci as master regulators of transposon activity in Drosophila. *Cell* 128, 1089-103 (2007).

656. Brennecke, J. et al. An epigenetic role for maternally inherited piRNAs in transposon silencing. *Science* 322, 1387-92 (2008).

657. Kunin, V., Sorek, R. and Hugenholtz, P. Evolutionary conservation of sequence and secondary structures in CRISPR repeats. *Genome Biol* 8, R61 (2007).

658. Bolotin, A., Quinquis, B., Sorokin, A. and Ehrlich, S.D. Clustered regularly interspaced short palindrome repeats (CRISPRs) have spacers of extrachromosomal origin. *Microbiology* 151, 2551-61 (2005).

659. Marraffini, L.A. and Sontheimer, E.J. Invasive DNA, chopped and in the CRISPR. *Structure* 17, 786-8 (2009).

660. Mojica, F.J., Diez-Villasenor, C., Garcia-Martinez, J. and Almendros, C. Short motif sequences determine the targets of the prokaryotic CRISPR defence system. *Microbiology* 155, 733-40 (2009).

661. Hale, C., Kleppe, K., Terns, R.M. and Terns, M.P. Prokaryotic silencing (psi)RNAs in Pyrococcus furiosus. *RNA* 14, 2572-9 (2008).

662. Hale, C.R. et al. RNA-guided RNA cleavage by a CRISPR RNA-Cas protein complex. *Cell* 139, 945-56 (2009).

663. Hale, C.J., Erhard, K.F., Jr., Lisch, D. and Hollick, J.B. Production and processing of siRNA precursor transcripts from the highly repetitive maize genome. *PLoS Genet* 5, e1000598 (2009).

664. Marraffini, L.A. and Sontheimer, E.J. CRISPR interference: RNA-directed adaptive immunity in bacteria and archaea. *Nat Rev Genet* 11, 181-190 (2010).

665. Horvath, P. et al. Diversity, activity, and evolution of CRISPR loci in Streptococcus thermophilus. *J Bacteriol* 190, 1401-12 (2008).

666. Luria, S.E. and Delbrück, M. Mutations of Bacteria from Virus Sensitivity to Virus Resistance. *Genetics* 28, 491-51 (1943).

667. Horvath, P. and Barrangou, R. CRISPR/Cas, the immune system of bacteria and archaea. *Science* 327, 167-70 (2010).

668. Jaskiewicz, M., Conrath, U. and Peterhansel, C. Chromatin modification acts as a memory for systemic acquired resistance in the plant stress response. *EMBO Rep* 12, 50-5 (2011).

669. Cairns, J., Overbaugh, J. and Miller, S. The origin of mutants. *Nature* 335, 142-5 (1988).

670. Maenhaut-Michel, G. and Shapiro, J.A. The roles of starvation and selective substrates in the emergence of araB-lacZ fusion clones. *Embo J* 13, 5229-39 (1994).

671. Weismann, A. *The Germ-Plasm: A Theory of Heredity*, (Charles Scribner's Sons, New York, 1893).

672. Bushman, F.D. Targeting survival: integration site selection by retroviruses and LTR-retrotransposons. *Cell* 115, 135-8 (2003).

673. Devine, S.E. and Boeke, J.D. Integration of the yeast retrotransposon Ty1 is targeted to regions upstream of genes transcribed by RNA polymerase III. *Genes Dev* 10, 620-33 (1996).

674. Bolton, E.C. and Boeke, J.D. Transcriptional interactions between yeast tRNA genes, flanking genes and Ty elements: a genomic point of view. *Genome Res* 13, 254-63 (2003).

675. Fauvarque, M.O. and Dura, J.M. polyhomeotic regulatory sequences induce developmental regulator-dependent variegation and targeted P-element insertions in Drosophila. *Genes Dev* 7, 1508-20 (1983).

676. Hama, C., Ali, Z. and Kornberg, T.B. Region-specific recombination and expression are directed by portions of the Drosophila engrailed promoter. *Genes Dev* 4, 1079-93 (1990).

677. Kassis, J.A., Noll, E., VanSickle, E.P., Odenwald, W.F. and Perrimon, N. Altering the insertional specificity of a Drosophila transposable element. *Proc Natl Acad Sci USA* 89, 1919-23 (1992).

678. Taillebourg, E. and Dura, J.M. A novel mechanism for P element homing in Drosophila. *Proc Natl Acad Sci USA* 96, 6856-61 (1999).

679. Bender, W. and Hudson, A. P element homing to the Drosophila bithorax complex. *Development* 127, 3981-92 (2000).

680. Kassis, J.A. Pairing-sensitive silencing, polycomb group response elements, and transposon homing in Drosophila. *Adv Genet* 46, 421-38 (2002).

681. Rubin, G.M. and Spradling, A.C. Genetic transformation of Drosophila with transposable element vectors. *Science* 218, 348-53 (1982).

682. Spradling, A.C. and Rubin, G.M. Transposition of cloned P elements into Drosophila germ line chromosomes. *Science* 218, 341-7 (1982).

683. Spradling, A.C. et al. Gene disruptions using P transposable elements: an integral component of the Drosophila genome project. *Proc Natl Acad Sci USA* 92, 10824-30 (1995).

684. Rubin, G.M. and Spradling, A.C. Vectors for P element-mediated gene transfer in Drosophila. *Nucleic Acids Res* 11, 6341-51 (1983).

685. Bateson, W. *Materials for the Study of Variation Treated With Especial Regard to Discontinuity in the Origin of Species* (Macmillan, London, 1894).

686. de Vries, H. *Species and Varieties, Their Origin by Mutation* (lectures delivered at the University of California). (Open Court Publishing Co., Chicago, 1905).

687. Goldschmidt, R. Sone aspects of evolution. *Science* 78, 539-47 (1933).

688. Margulis, L. Symbiosis and evolution. *Sci Am* 225, 48-57 (1971).

689. Levy, S.B. The challenge of antibiotic resistance. *Sci Am* 278, 46-53 (1998).

690. Gorini, L. Antibiotics and the Genetic Code. *Sci Am* 214, 102-109 (1966).

691. Ryan, F.J. Evolution observed. *Sci Am* 189, 78-82 (1953).

692. Fauci, A.S. Infectious diseases: considerations for the 21st century. *Clin Infect Dis* 32, 675-85 (2001).

693. Watanabe, T. and Fukasawa, T. Episome-mediated transfer of drug resistance in Enterobacteriaceae. III. Transduction of resistance factors. *J Bacteriol* 82, 202-9 (1961).

694. Watanabe, T. Episome-Mediated Transfer of Drug Resistance in Enterobacteriaceae. Vi. High-Frequency Resistance Transfer System in Escherichia Coli. *J Bacteriol* 85, 788-94 (1963).

695. Watanabe, T. Infectious drug resistance. *Sci Am* 217, 19-28 (1967).

696. Benveniste, R. and Davies, J. Mechanisms of antibiotic resistance in bacteria. *Annu Rev Biochem* 42, 471-506 (1973).

697. Foster, T.J. Plasmid-determined resistance to antimicrobial drugs and toxic metal ions in bacteria. *Microbiol Rev* 47, 361-409 (1983).

698. Mazel, D. and Davies, J. Antibiotic resistance in microbes. *Cell Mol Life* Sci 56, 742-54 (1999).

699. Iyer, L.M., Balaji, S., Koonin, E.V. and Aravind, L. Evolutionary genomics of nucleo-cytoplasmic large DNA viruses. *Virus Res* 117, 156-84 (2006).

700. Filee, J., Pouget, N. and Chandler, M. Phylogenetic evidence for extensive lateral acquisition of cellular genes by Nucleocytoplasmic large DNA viruses. *BMC Evol Biol* 8, 320 (2008).

701. Yutin, N., Wolf, Y.I., Raoult, D. and Koonin, E.V. Eukaryotic large nucleo-cytoplasmic DNA viruses: clusters of orthologous genes and reconstruction of viral genome evolution. *Virol J* 6, 223 (2009).

702. Yutin, N. and Koonin, E.V. Evolution of DNA ligases of Nucleo-Cytoplasmic Large DNA viruses of eukaryotes: a case of hidden complexity. *Biol Direct* 4, 51 (2009).

703. Hall, R.M. and Collis, C.M. Mobile gene cassettes and integrons: capture and spread of genes by site-specific recombination. *Mol Microbiol* 15, 593-600 (1995).

704. Rowe-Magnus, D.A. and Mazel, D. The role of integrons in antibiotic resistance gene capture. *Int J Med Microbiol* 292, 115-25 (2002).

705. Labbate, M., Case, R.J. and Stokes, H.W. The integron/gene cassette system: an active player in bacterial adaptation. *Methods Mol Biol* 532, 103-25 (2009).

706. Bouvier, M., Ducos-Galand, M., Loot, C., Bikard, D. and Mazel, D. Structural features of single-stranded integron cassette attC sites and their role in strand selection. *PLoS Genet* 5, e1000632 (2009).

707. Fluit, A.C. and Schmitz, F.J. Resistance integrons and super-integrons. *Clin Microbiol Infect* 10, 272-88 (2004).

708. Vaisvila, R., Morgan, R.D., Posfai, J. and Raleigh, E.A. Discovery and distribution of super-integrons among pseudomonads. *Mol Microbiol* 42, 587-601 (2001).

709. Rowe-Magnus, D.A., Guérout, A.M. and Mazel, D. Super-integrons. *Res Microbiol* 150, 641-51 (1999).

710. Lee, C.A., Babic, A. and Grossman, A.D. Autonomous plasmid-like replication of a conjugative transposon. *Mol Microbiol* 75, 268-79 (2010).

711. Burrus, V., Pavlovic, G., Decaris, B. and Guedon, G. Conjugative transposons: the tip of the iceberg. *Mol Microbiol* 46, 601-10 (2002).

712. Sonea, S. and Panisset, M. *A New Bacteriology*, (Jones and Batlett, Boston, 1983).

713. Sonea, S. Bacterial plasmids instrumental in the origin of eukaryotes? *Rev Can Biol* 31, 61-3 (1972).

714. Sonea, S. A bacterial way of life. *Nature* 331, 216 (1988).

715. van der Meer, J.R. and Sentchilo, V. Genomic islands and the evolution of catabolic pathways in bacteria. *Curr Opin Biotechnol* 14, 248-54 (2003).

716. Tallent, S.M., Langston, T.B., Moran, R.G. and Christie, G.E. Transducing particles of Staphylococcus aureus pathogenicity island SaPI1 are comprised of helper phage-encoded proteins. *J Bacteriol* 189, 7520-4 (2007).

717. Tormo, M.A. et al. Staphylococcus aureus pathogenicity island DNA is packaged in particles composed of phage proteins. *J Bacteriol* 190, 2434-40 (2008).

718. Tormo-Mas, M.A. et al. Moonlighting bacteriophage proteins derepress staphylococcal pathogenicity islands. *Nature* 465, 779-82 (2010).

719. Reed, C. Sequencing Sea World. *Sci Am* 295, 23-24 (2006).

720. Tyson, G.W. et al. Community structure and metabolism through reconstruction of microbial genomes from the environment. *Nature* 428, 37-43 (2004).

721. Venter, J.C. et al. Environmental genome shotgun sequencing of the Sargasso Sea. *Science* 304, 66-74 (2004).

722. Ventura, M. et al. Microbial diversity in the human intestine and novel insights from metagenomics. *Front Biosci* 14, 3214-21 (2009).

723. Wooley, J.C., Godzik, A. and Friedberg, I. A primer on metagenomics. *PLoS Comput Biol* 6, e1000667 (2010).

724. Ellrott, K., Jaroszewski, L., Li, W., Wooley, J.C. and Godzik, A. Expansion of the protein repertoire in newly explored environments: human gut microbiome specific protein families. *PLoS Comput Biol* 6, e1000798 (2010).

725. Jones, B.V., Sun, F. and Marchesi, J.R. Comparative metagenomic analysis of plasmid encoded functions in the human gut microbiome. *BMC Genomics* 11, 46 (2010).

726. Qin, J. et al. A human gut microbial gene catalogue established by metagenomic sequencing. *Nature* 464, 59-65 (2010).

727. Nelson, K.E. et al. A catalog of reference genomes from the human microbiome. *Science* 328, 994-9 (2010).

728. Achtman, M. and Wagner, M. Microbial diversity and the genetic nature of microbial species. *Nat Rev Microbiol* 6, 431-40 (2008).

729. Schloss, P.D. and Handelsman, J. Metagenomics for studying unculturable microorganisms: cutting the Gordian knot. *Genome Biol* 6, 229 (2005).

730. Handelsman, J. Metagenomics: application of genomics to uncultured microorganisms. *Microbiol Mol Biol Rev* 68, 669-85 (2004).

731. Edwards, R.A. and Rohwer, F. Viral metagenomics. *Nat Rev Microbiol* 3, 504-10 (2005).

732. Bench, S.R. et al. Metagenomic characterization of Chesapeake Bay virioplankton. *Appl Environ Microbiol* 73, 7629-41 (2007).

733. Sandaa, R.A., Clokie, M. and Mann, N.H. Photosynthetic genes in viral populations with a large genomic size range from Norwegian coastal waters. *FEMS Microbiol Ecol* 63, 2-11 (2008).

734. Yanofsky, C. Gene Structure and Protein Structure. *Sci Am* 216, 80-94 (1967).

735. Wilson, A.C. The Molecular Basis of Evolution. *Sci Am* 253, 164-173 (1985).

736. Doolittle, R.F. and Bork, P. Evolutionarily mobile modules in proteins. *Sci Am* 269, 50-6 (1993).

737. Bjorklund, A.K., Ekman, D. and Elofsson, A. Expansion of protein domain repeats. *PLoS Comput Biol* 2, e114 (2006).

738. Bjorklund, A.K., Ekman, D., Light, S., Frey-Skott, J. and Elofsson, A. Domain rearrangements in protein evolution. *J Mol Biol* 353, 911-23 (2005).

739. Schmidt, E.E. and Davies, C.J. The origins of polypeptide domains. *Bioessays* 29, 262-70 (2007).

740. Doolittle, R.F. The Roots of Bioinformatics in Protein Evolution. *PLoS Comput Biol* 6, e1000875 (2010).

741. Pabo, C.O., Sauer, R.T., Sturtevant, J.M. and Ptashne, M. The lambda repressor contains two domains. *Proc Natl Acad Sci USA* 76, 1608-12 (1979).

742. McKnight, S.L. Molecular Zippers in Gene Regulation. *Sci Am* 264, 54-64 (1991).

743. Lewis, M. The lac repressor. *C R Biol* 328, 521-48 (2005).

744. Marsden, R.L. et al. Exploiting protein structure data to explore the evolution of protein function and biological complexity. *Philos Trans R Soc Lond B Biol Sci* 361, 425-40 (2006).

745. Gonzalez, M.W. and Pearson, W.R. RefProtDom: A Protein Database with Improved Domain Boundaries and Homology Relationships. *Bioinformatics* (2010).

746. Kolkman, J.A. and Stemmer, W.P. Directed evolution of proteins by exon shuf-fling. *Nat Biotechnol* 19, 423-8 (2001).

747. Miller, J.H. et al. Fusions of the lac and trp Regions of the Escherichia coli Chromosome. *J Bacteriol* 104, 1273-9 (1970).

748. Casadaban, M.J. Transposition and fusion of the lac genes to selected promot-ers in Escherichia coli using bacteriophage lambda and Mu. *J Mol Biol* 104, 541-55 (1976).

749. Moran, J.V. Human L1 retrotransposition: insights and peculiarities learned from a cultured cell retrotransposition assay. *Genetica* 107, 39-51 (1999).

750. Xing, J. et al. Emergence of primate genes by retrotransposon-mediated sequence transduction. *Proc Natl Acad Sci USA* 103, 17608-13 (2006).

751. Lisch, D. Pack-MULEs: theft on a massive scale. *Bioessays* 27, 353-5 (2005).

752. Lai, J., Li, Y., Messing, J. and Dooner, H.K. Gene movement by Helitron transposons contributes to the haplotype variability of maize. *Proc Natl Acad Sci USA* 102, 9068-73 (2005).

753. Lal, S.K. and Hannah, L.C. Helitrons contribute to the lack of gene colinear-ity observed in modern maize inbreds. *Proc Natl Acad Sci USA* 102, 9993-4 (2005).

754. Lal, S.K. and Hannah, L.C. Plant genomes: massive changes of the maize genome are caused by Helitrons. *Heredity* 95, 421-2 (2005).

755. Young, J.M. and Trask, B.J. The sense of smell: genomics of vertebrate odor-ant receptors. *Hum Mol Genet* 11, 1153-60 (2002).

756. Tatusov, R.L., Koonin, E.V. and Lipman, D.J. A genomic perspective on pro-tein families. *Science* 278, 631-7 (1997).

757. Harrison, P.M. and Gerstein, M. Studying genomes through the aeons: protein families, pseudogenes and proteome evolution. *J Mol Biol* 318, 1155-74 (2002).

758. Lespinet, O., Wolf, Y.I., Koonin, E.V. and Aravind, L. The role of lineage-specific gene family expansion in the evolution of eukaryotes. *Genome Res* 12, 1048-59 (2002).

759. Jordan, I.K., Makarova, K.S., Spouge, J.L., Wolf, Y.I. and Koonin, E.V. Lineage-specific gene expansions in bacterial and archaeal genomes. *Genome Res* 11, 555-65 (2001).

760. Pushker, R., Mira, A. and Rodriguez-Valera, F. Comparative genomics of gene-family size in closely related bacteria. *Genome Biol* 5, R27 (2004).

761. Gordon, S.V. et al. Genomics of Mycobacterium bovis. *Tuberculosis (Edinb)* 81, 157-63 (2001).

762. Brennan, M.J. and Delogu, G. The PE multigene family: a "molecular mantra" for mycobacteria. *Trends Microbiol* 10, 246-9 (2002).

763. Axel, R. The Molecular Logic Of Smell. *Sci Am* 295, 68-75 (2006).

764. Crasto, C., Singer, M.S. and Shepherd, G.M. The olfactory receptor family album. *Genome Biol* 2, REVIEWS1027 (2001).

765. Niimura, Y. and Nei, M. Evolutionary dynamics of olfactory and other chemosensory receptor genes in vertebrates. *J Hum Genet* 51, 505-17 (2006).

766. Young, J.M. et al. Different evolutionary processes shaped the mouse and human olfactory receptor gene families. *Hum Mol Genet* 11, 535-46 (2002).

767. Sogin, S.J., Sogin, M.L. and Woese, C.R. Phylogenetic measurement in procaryotes by primary structural characterization. *J Mol Evol* 1, 173-84 (1971).

768. Lake, J.A. The Ribosome. *Sci Am* 245, 84-97 (1981).

769. Woese, C.R. Archaebacteria. *Sci Am* 244, 98-122 (1981).

770. Balch, W.E., Magrum, L.J., Fox, G.E., Wolfe, R.S. and Woese, C.R. An ancient divergence among the bacteria. *J Mol Evol* 9, 305-11 (1977).

771. Woese, C.R. and Fox, G.E. Phylogenetic structure of the prokaryotic domain: the primary kingdoms. *Proc Natl Acad Sci USA* 74, 5088-90 (1977).

772. Woese, C.R., Magrum, L.J. and Fox, G.E. Archaebacteria. *J Mol Evol* 11, 245-51 (1978).

773. Balch, W.E., Fox, G.E., Magrum, L.J., Woese, C.R. and Wolfe, R.S. Methanogens: reevaluation of a unique biological group. *Microbiol Rev* 43, 260-96 (1979).

774. Woese, C.R. A proposal concerning the origin of life on the planet earth. *J Mol Evol* 13, 95-101 (1979).

775. DeLong, E.F. Everything in moderation: archaea as "non-extremophiles." *Curr Opin Genet Dev* 8, 649-54 (1998).

776. Woese, C.R., Kandler, O. and Wheelis, M.L. Towards a natural system of organisms: proposal for the domains Archaea, Bacteria, and Eucarya. *Proc Natl Acad Sci USA* 87, 4576-9 (1990).

777. Raymond, J. The role of horizontal gene transfer in photosynthesis, oxygen production, and oxygen tolerance. *Methods Mol Biol* 532, 323-38 (2009).

778. Aravind, L., Tatusov, R.L., Wolf, Y.I., Walker, D.R. and Koonin, E.V. Evidence for massive gene exchange between archaeal and bacterial hyperthermophiles. *Trends Genet* 14, 442-4 (1998).

779. Kyrpides, N.C. and Olsen, G.J. Archaeal and bacterial hyperthermophiles: horizontal gene exchange or common ancestry? *Trends Genet* 15, 298-9 (1999).

780. Pereira, S.L. and Reeve, J.N. Histones and nucleosomes in Archaea and Eukarya: a comparative analysis. *Extremophiles* 2, 141-8 (1998).

781. Robinson, N.P. and Bell, S.D. Origins of DNA replication in the three domains of life. *FEBS J* 272, 3757-66 (2005).

782. Hickey, A.J., Conway de Macario, E. and Macario, A.J. Transcription in the archaea: basal factors, regulation, and stress-gene expression. *Crit Rev Biochem Mol Biol* 37, 537-99 (2002).

783. Horiike, T., Hamada, K. and Shinozawa, T. Origin of eukaryotic cell nuclei by symbiosis of Archaea in Bacteria supported by the newly clarified origin of functional genes. *Genes Genet Syst* 77, 369-76 (2002).

784. Horiike, T., Hamada, K., Kanaya, S. and Shinozawa, T. Origin of eukaryotic cell nuclei by symbiosis of Archaea in Bacteria is revealed by homology-hit analysis. *Nat Cell Biol* 3, 210-4 (2001).

785. Martin, W. Archaebacteria (Archaea) and the origin of the eukaryotic nucleus. *Curr Opin Microbiol* 8, 630-7 (2005).

786. Yutin, N., Makarova, K.S., Mekhedov, S.L., Wolf, Y.I. and Koonin, E.V. The deep archaeal roots of eukaryotes. *Mol Biol Evol* 25, 1619-30 (2008).

787. Cox, C.J., Foster, P.G., Hirt, R.P., Harris, S.R. and Embley, T.M. The archae-bacterial origin of eukaryotes. *Proc Natl Acad Sci USA* 105, 20356-61 (2008).

788. Fournier, G.P., Huang, J. and Gogarten, J.P. Horizontal gene transfer from extinct and extant lineages: biological innovation and the coral of life. *Philos Trans R Soc Lond B Biol Sci* 364, 2229-39 (2009).

789. Huang, J. and Gogarten, J.P. Ancient gene transfer as a tool in phylogenetic reconstruction. *Methods Mol Biol* 532, 127-39 (2009).

790. Honegger, R. Simon Schwendener (1829–1919) and the dual hypothesis in lichens. *Bryologist* 103, 307-13 (2000).

791. Sapp, J. *Evolution by Association: A History of Symbiosis*, (Oxford University Press, Oxford, 1994).

792. Sapp, J., Carrapico, F. and Zolotonosov, M. Symbiogenesis: the hidden face of Constantin Merezhkowsky. *Hist Philos Life Sci* 24, 413-40 (2002).

793. Kozo-Polyansky, B.M. *Symbiogenesis: A New Principle of Evolution (1924)*, (Harvard University Press, Cambridge, MA, 2010).

794. Walin, I.E. *Symbionticism and the Origin of Species*, (Williams and Wilkins, Baltimore, 1927).

795. Dippell, R.V. Mutations of the killer plasmagene, kappa, in variety 4 of Paramecium aurelia. *Am Nat* 82, 43-50 (1948).

796. Sonneborn, T.M. Gene and Cytoplasm: I. The Determination and Inheritance of the Killer Character in Variety 4 of Paramecium Aurelia. *Proc Natl Acad Sci USA* 29, 329-38 (1943).

797. Sonneborn, T.M. Cellular development and heredity. *J Indiana State Med Assoc* 60, 1036-8 (1967).

798. Gibson, I. The endosymbionts of Paramecium. *CRC Crit Rev Microbiol* 3, 243-73 (1974).

799. Preer, J.R., Jr., Preer, L.B. and Jurand, A. Kappa and other endosymbionts in Paramecium aurelia. *Bacteriol Rev* 38, 113-63 (1974).

800. Engelstadter, J. and Telschow, A. Cytoplasmic incompatibility and host population structure. *Heredity* 103, 196-207 (2009).

801. Sharon, G. et al. Commensal bacteria play a role in mating preference of Drosophila melanogaster. *Proc Natl Acad Sci USA* 107, 20,051-6 (2010).

802. Stams, A.J. Metabolic interactions between anaerobic bacteria in methanogenic environments. *Antonie Van Leeuwenhoek* 66, 271-94 (1994).

803. Stams, A.J., Oude Elferink, S.J. and Westermann, P. Metabolic interactions between methanogenic consortia and anaerobic respiring bacteria. *Adv Biochem Eng Biotechnol* 81, 31-56 (2003).

804. Stams, A.J. and Plugge, C.M. Electron transfer in syntrophic communities of anaerobic bacteria and archaea. *Nat Rev Microbiol* 7, 568-77 (2009).

805. Paerl, H.W. and Pinckney, J.L. A Mini-review of Microbial Consortia: Their Roles in Aquatic Production and Biogeochemical Cycling. *Microb Ecol* 31, 225-47 (1996).

806. Froestl, J.M. and Overman, J. Phylogenetic affiliation of the bacteria that constitute phototrophic consortia. *Arch Microbiol* 174, 50-58 (2000).

807. Wanner, G., Vogl, K. and Overmann, J. Ultrastructural characterization of the prokaryotic symbiosis in "Chlorochromatium aggregatum." *J Bacteriol* 190, 3721-30 (2008).

808. Kanzler, B.E., Pfannes, K.R., Vogl, K. and Overmann, J. Molecular characterization of the nonphotosynthetic partner bacterium in the consortium "Chlorochromatium aggregatum." *Appl Environ Microbiol* 71, 7434-41 (2005).

809. Martin, F., Kohler, A. and Duplessis, S. Living in harmony in the wood underground: ectomycorrhizal genomics. *Curr Opin Plant Biol* 10, 204-10 (2007).

810. Martin, F. and Nehls, U. Harnessing ectomycorrhizal genomics for ecological insights. *Curr Opin Plant Biol* 12, 508-15 (2009).

811. Delano-Frier, J.P. and Tejeda-Sartorius, M. Unraveling the network: Novel developments in the understanding of signaling and nutrient exchange mechanisms in the arbuscular mycorrhizal symbiosis. *Plant Signal Behav* 3, 936-44 (2008).

812. Teixeira, L., Ferreira, A. and Ashburner, M. The bacterial symbiont Wolbachia induces resistance to RNA viral infections in Drosophila melanogaster. *PLoS Biol* 6, e2 (2008).

813. Zientz, E., Feldhaar, H., Stoll, S. and Gross, R. Insights into the microbial world associated with ants. *Arch Microbiol* 184, 199-206 (2005).

814. Zientz, E., Dandekar, T. and Gross, R. Metabolic interdependence of obligate intracellular bacteria and their insect hosts. *Microbiol Mol Biol Rev* 68, 745-70 (2004).

815. de Souza, D.J., Bezier, A., Depoix, D., Drezen, J.M. and Lenoir, A. Blochmannia endosymbionts improve colony growth and immune defence in the ant Camponotus fellah. *BMC Microbiol* 9, 29 (2009).

816. Sun, S. and Cline, T.W. Effects of Wolbachia infection and ovarian tumor mutations on Sex-lethal germline functioning in Drosophila. *Genetics* 181, 1291-301 (2009).

817. Sacchi, L. et al. Bacteriocyte-like cells harbour Wolbachia in the ovary of Drosophila melanogaster (Insecta, Diptera) and Zyginidia pullula (Insecta, Hemiptera). *Tissue Cell* 42, 328-33 (2010).

818. Buchner, P. *Endosymbiosis of Animals with Plant Microorganisms*, (John Wiley and Sons, Chichester, 1965).

819. Heddi, A. et al. Molecular and cellular profiles of insect bacteriocytes: mutualism and harm at the initial evolutionary step of symbiogenesis. *Cell Microbiol* 7, 293-305 (2005).

820. Feldhaar, H. and Gross, R. Insects as hosts for mutualistic bacteria. *Int J Med Microbiol* 299, 1-8 (2009).

821. McFall-Ngai, M. Host-microbe symbiosis: the squid-Vibrio association—a naturally occurring, experimental model of animal/bacterial partnerships. *Adv Exp Med Biol* 635, 102-12 (2008).

822. McFall-Ngai, M.J. Negotiations between animals and bacteria: the "diplomacy" of the squid-vibrio symbiosis. *Comp Biochem Physiol A Mol Integr Physiol* 126, 471-80 (2000).

823. Baumann, P. Biology bacteriocyte-associated endosymbionts of plant sap-sucking insects. *Annu Rev Microbiol* 59, 155-89 (2005).

824. Douglas, A.E. Nutritional interactions in insect-microbial symbioses: aphids and their symbiotic bacteria Buchnera. *Annu Rev Entomol* 43, 17-37 (1998).

825. Narita, S., Kageyama, D., Nomura, M. and Fukatsu, T. Unexpected mechanism of symbiont-induced reversal of insect sex: feminizing Wolbachia continuously acts on the butterfly Eurema hecabe during larval development. *Appl Environ Microbiol* 73, 4332-41 (2007).

826. Gehrig, H., Schussler, A. and Kluge, M. Geosiphon pyriforme, a fungus forming endocytobiosis with Nostoc (cyanobacteria), is an ancestral member of the Glomales: evidence by SSU rRNA analysis. *J Mol Evol* 43, 71-81 (1996).

827. Queller, D.C. and Strassmann, J.E. Beyond society: the evolution of organismality. *Philos Trans R Soc Lond B Biol Sci* 364, 3143-55 (2009).

828. Goodenough, U.W. and Levine, R.P. The genetic activity of mitochondria and chloroplasts. *Sci Am* 223, 22-29 (1970).

829. Bonen, L. and Doolittle, W.F. On the prokaryotic nature of red algal chloroplasts. *Proc Natl Acad Sci USA* 72, 2310-4 (1975).

830. Zablen, L.B., Kissil, M.S., Woese, C.R. and Buetow, D.E. Phylogenetic origin of the chloroplast and prokaryotic nature of its ribosomal RNA. *Proc Natl Acad Sci USA* 72, 2418-22 (1975).

831. Woese, C.R. Endosymbionts and mitochondrial origins. *J Mol Evol* 10, 93-6 (1977).

832. Gray, M.W., Burger, G. and Lang, B.F. Mitochondrial evolution. *Science* 283, 1476-81 (1999).

833. Nozaki, H. A new scenario of plastid evolution: plastid primary endosymbiosis before the divergence of the "Plantae," emended. *J Plant Res* 118, 247-55 (2005).

834. Archibald, J.M.K., P. J. Recycled plastids: a "green movement" in eukaryotic evolution. *Trends Genet* 18, 577-584 (2002).

835. Lane, C.E. and Archibald, J.M. The eukaryotic tree of life: endosymbiosis takes its TOL. *Trends Ecol Evol* 23, 268-75 (2008).

836. Bhattacharya, D., Yoon, H.S. and Hackett, J.D. Photosynthetic eukaryotes unite: endosymbiosis connects the dots. *Bioessays* 26, 50-60 (2004).

837. Reyes-Prieto, A., Weber, A.P. and Bhattacharya, D. The origin and establishment of the plastid in algae and plants. *Annu Rev Genet* 41, 147-68 (2007).

838. Martin, W. et al. Early cell evolution, eukaryotes, anoxia, sulfide, oxygen, fungi first (?), and a tree of genomes revisited. *IUBMB Life* 55, 193-204 (2003).

839. Margulis, L., Chapman, M., Guerrero, R. and Hall, J. The last eukaryotic common ancestor (LECA): acquisition of cytoskeletal motility from aerotolerant spirochetes in the Proterozoic Eon. *Proc Natl Acad Sci USA* 103, 13080-5 (2006).

840. Embley, T.M. Multiple secondary origins of the anaerobic lifestyle in eukaryotes. *Philos Trans R Soc Lond B Biol Sci* 361, 1055-67 (2006).

841. Embley, T.M. and Martin, W. Eukaryotic evolution, changes and challenges. *Nature* 440, 623-30 (2006).

842. Grivell, L.A. Mitochondrial DNA. *Sci Am* 248, 78-89 (1983).

843. Gray, M.W., Burger, G. and Lang, B.F. The origin and early evolution of mitochondria. *Genome Biol* 2, REVIEWS1018 (2001).

844. Lang, B.F., Gray, M.W. and Burger, G. Mitochondrial genome evolution and the origin of eukaryotes. *Annu Rev Genet* 33, 351-97 (1999).

845. Burger, G., Gray, M.W. and Lang, B.F. Mitochondrial genomes: anything goes. *Trends Genet* 19, 709-16 (2003).

846. Archibald, J.M., Rogers, M.B., Toop, M., Ishida, K. and Keeling, P.J. Lateral gene transfer and the evolution of plastid-targeted proteins in the secondary plastid-containing alga Bigelowiella natans. *Proc Natl Acad Sci USA* 100, 7678-83 (2003).

847. Huang, J. et al. Phylogenomic evidence supports past endosymbiosis, intracellular and horizontal gene transfer in Cryptosporidium parvum. *Genome Biol* 5, R88 (2004).

848. Bock, R. and Timmis, J.N. Reconstructing evolution: gene transfer from plastids to the nucleus. *Bioessays* 30, 556-66 (2008).

849. Gray, M.W., Lang, B.F. and Burger, G. Mitochondria of protists. *Annu Rev Genet* 38, 477-524 (2004).

850. Westenberger, S.J. et al. Trypanosoma cruzi mitochondrial maxicircles display species- and strain-specific variation and a conserved element in the noncoding region. *BMC Genomics* 7, 60 (2006).

851. Lukes, J., Hashimi, H. and Zikova, A. Unexplained complexity of the mitochondrial genome and transcriptome in kinetoplastid flagellates. *Curr Genet* 48, 277-99 (2005).

852. Stuart, K., Allen, T.E., Heidmann, S. and Seiwert, S.D. RNA editing in kinetoplastid protozoa. *Microbiol Mol Biol Rev* 61, 105-20 (1997).

853. Byrne, E.M., Connell, G.J. and Simpson, L. Guide RNA-directed uridine insertion RNA editing in vitro. *Embo J* 15, 6758-65 (1996).

854. Landweber, L.F. and Gilbert, W. Phylogenetic analysis of RNA editing: a primitive genetic phenomenon. *Proc Natl Acad Sci USA* 91, 918-21 (1994).

855. Thomas, S., Martinez, L.L., Westenberger, S.J. and Sturm, N.R. A population study of the minicircles in Trypanosoma cruzi: predicting guide RNAs in the absence of empirical RNA editing. *BMC Genomics* 8, 133 (2007).

856. Keeling, P.J. Chromalveolates and the evolution of plastids by secondary endosymbiosis. *J Eukaryot Microbiol* 56, 1-8 (2009).

857. Gould, S.B., Waller, R.F. and McFadden, G.I. Plastid evolution. *Annu Rev Plant Biol* 59, 491-517 (2008).

858. Deschamps, P. et al. Metabolic symbiosis and the birth of the plant kingdom. *Mol Biol Evol* 25, 536-48 (2008).

859. Mittag, M. Circadian rhythms in microalgae. *Int Rev Cytol* 206, 213-47 (2001).

860. Hader, D.P. Gravitaxis in unicellular microorganisms. *Adv Space Res* 24, 843-50 (1999).

861. Ginger, M.L. Trypanosomatid biology and euglenozoan evolution: new insights and shifting paradigms revealed through genome sequencing. *Protist* 156, 377-92 (2005).

862. Hader, D.P. and Lebert, M. Photoorientation in photosynthetic flagellates. *Methods Mol Biol* 571, 51-65 (2009).

863. Goto, K. and Beneragama, C.K. Circadian clocks and antiaging: do non-aging microalgae like Euglena reveal anything? *Ageing Res Rev* 9, 91-100 (2010).

864. Leander, B.S., Esson, H.J. and Breglia, S.A. Macroevolution of complex cytoskeletal systems in euglenids. *Bioessays* 29, 987-1000 (2007).

865. Cavalier-Smith, T. The phagotrophic origin of eukaryotes and phylogenetic classification of Protozoa. *Int J Syst Evol Microbiol* 52, 297-354 (2002).

866. Archibald, J.M. Nucleomorph genomes: structure, function, origin and evolution. *Bioessays* 29, 392-402 (2007).

867. Silver, T.D. et al. Phylogeny and nucleomorph karyotype diversity of chlorarachniophyte algae. *J Eukaryot Microbiol* 54, 403-10 (2007).

868. Cavalier-Smith, T. Kingdom protozoa and its 18 phyla. *Microbiol Rev* 57, 953-94 (1993).

869. Roberts, K., Granum, E., Leegood, R.C. and Raven, J.A. Carbon acquisition by diatoms. *Photosynth Res* 93, 79-88 (2007).

870. Nisbet, R.E., Kilian, O. and McFadden, G.I. Diatom genomics: genetic acquisitions and mergers. *Curr Biol* 14, R1048-50 (2004).

871. Armbrust, E.V. et al. The genome of the diatom Thalassiosira pseudonana: ecology, evolution, and metabolism. *Science* 306, 79-86 (2004).

872. Gould, S.B. et al. Nucleus-to-nucleus gene transfer and protein retargeting into a remnant cytoplasm of cryptophytes and diatoms. *Mol Biol Evol* 23, 2413-22 (2006).

873. Venn, A.A., Loram, J.E. and Douglas, A.E. Photosynthetic symbioses in animals. *J Exp Bot* 59, 1069-80 (2008).

874. Martin, W. and Muller, M. The hydrogen hypothesis for the first eukaryote. *Nature* 392, 37-41 (1998).

875. Gross, J. and Bhattacharya, D. Uniting sex and eukaryote origins in an emerging oxygenic world. *Biol Direct* 5, 53 (2010).

876. Lake, J.A. Evidence for an early prokaryotic endosymbiosis. *Nature* 460, 967-71 (2009).

877. Rivera, M.C.L., J. A. . The ring of life: evidence for a genome fusion origin of eukaryotes. *Nature* 431, 152-155 (2004).

878. Dagan, T., Roettger, M., Bryant, D. and Martin, W. Genome networks root the tree of life between prokaryotic domains. *Genome Biol Evol* 2, 379-92 (2010).

879. Margulis, L. *Symbiosis in Cell Evolution*, (W.H. Freeman Co, London, 1981).

880. Dolan, M.F., Melnitsky, H., Margulis, L. and Kolnicki, R. Motility proteins and the origin of the nucleus. *Anat Rec* 268, 290-301 (2002).

881. Chapman, M.J., Dolan, M.F. and Margulis, L. Centrioles and kinetosomes: form, function, and evolution. *Q Rev Biol* 75, 409-29 (2000).

882. Margulis, L., Dolan, M.F. and Guerrero, R. The chimeric eukaryote: origin of the nucleus from the karyomastigont in amitochondriate protists. *Proc Natl Acad Sci USA* 97, 6954-9 (2000).

883. Wier, A.M. et al. Spirochete attachment ultrastructure: Implications for the origin and evolution of cilia. *Biol Bull* 218, 25-35 (2010).

884. Carvalho-Santos, Z. et al. Stepwise evolution of the centriole-assembly pathway. *J Cell Sci* 123, 1414-26 (2010).

885. Williamson, D. *Larvae and Evolution: Toward a New Zoology*, (Chapman and Hall, New York, 1992).

886. Williamson, D.I. *The Origins of Larvae*, (Kluwer, Dordrecht, 2003).

887. Williamson, D.I. Caterpillars evolved from onychophorans by hybridogenesis. *Proc Natl Acad Sci USA* 106, 19901-5 (2009).

888. Epel, D. The Program of Fertilization. *Sci Am* 237, 128-138 (1977).

889. Wassarman, P.M. Fertilization in Mammals. *Sci Am* 259, 78-84 (1988).

890. Schatten, H. and Sun, Q.Y. The role of centrosomes in mammalian fertilization and its significance for ICSI. *Mol Hum Reprod* 15, 531-8 (2009).

891. Oldroyd, G.E., Harrison, M.J. and Paszkowski, U. Reprogramming plant cells for endosymbiosis. *Science* 324, 753-4 (2009).

892. Jones, K.M., Kobayashi, H., Davies, B.W., Taga, M.E. and Walker, G.C. How rhizobial symbionts invade plants: the Sinorhizobium-Medicago model. *Nat Rev Microbiol* 5, 619-33 (2007).

893. Wang, D. et al. A nodule-specific protein secretory pathway required for nitrogen-fixing symbiosis. *Science* 327, 1126-9 (2010).

894. Van de Velde, W. et al. Plant peptides govern terminal differentiation of bacteria in symbiosis. *Science* 327, 1122-6 (2010).

895. Cao, H. et al. Complex quorum-sensing regulatory systems regulate bacterial growth and symbiotic nodulation in Mesorhizobium tianshanense. *Arch Microbiol* 191, 283-9 (2009).

896. Deakin, W.J. and Broughton, W.J. Symbiotic use of pathogenic strategies: rhizobial protein secretion systems. *Nat Rev Microbiol* 7, 312-20 (2009).

897. Chesnick, J.M. and Cox, E.R. Synchronized sexuality of an algal symbiont and its dinoflagellate host, Peridinium balticum (Levander) Lemmermann. *Biosystems* 21, 69-78 (1987).

898. O'Donnell, A.J., Schneider, P., McWatters, H.G. and Reece, S.E. Fitness costs of disrupting circadian rhythms in malaria parasites. *Proceedings of the Royal Society B: Biological Sciences* (2011).

899. Gramzow, L., Ritz, M.S. and Theissen, G. On the origin of MADS-domain transcription factors. *Trends Genet* 26, 149-53 (2010).

900. Larroux, C. et al. Genesis and expansion of metazoan transcription factor gene classes. *Mol Biol Evol* 25, 980-96 (2008).

901. Derelle, R., Lopez, P., Le Guyader, H. and Manuel, M. Homeodomain proteins belong to the ancestral molecular toolkit of eukaryotes. *Evol Dev* 9, 212-9 (2007).

902. Piriyapongsa, J., Rutledge, M.T., Patel, S., Borodovsky, M. and Jordan, I.K. Evaluating the protein coding potential of exonized transposable element sequences. *Biol Direct* 2, 31 (2007).

903. Piriyapongsa, J., Polavarapu, N., Borodovsky, M. and McDonald, J. Exonization of the LTR transposable elements in human genome. *BMC Genomics* 8, 291 (2007).

904. Corvelo, A. and Eyras, E. Exon creation and establishment in human genes. *Genome Biol* 9, R141 (2008).

905. Sorek, R. The birth of new exons: mechanisms and evolutionary consequences. *RNA* 13, 1603-8 (2007).

906. Schwartz, S. et al. Alu exonization events reveal features required for precise recognition of exons by the splicing machinery. *PLoS Comput Biol* 5, e1000300 (2009).

907. Nekrutenko, A. and Li, W.H. Transposable elements are found in a large number of human protein-coding genes. *Trends Genet* 17, 619-21 (2001).

908. Sela, N. et al. Comparative analysis of transposed element insertion within human and mouse genomes reveals Alu's unique role in shaping the human transcriptome. *Genome Biol* 8, R127 (2007).

909. Sela, N., Mersch, B., Hotz-Wagenblatt, A. and Ast, G. Characteristics of transposable element exonization within human and mouse. *PLoS One* 5, e10907 (2010).

910. Knerr, I. et al. Endogenous retroviral syncytin: compilation of experimental research on syncytin and its possible role in normal and disturbed human placentogenesis. *Mol Hum Reprod* 10, 581-8 (2004).

911. Ono, R. et al. Deletion of Peg10, an imprinted gene acquired from a retrotransposon, causes early embryonic lethality. *Nat Genet* 38, 101-6 (2006).

912. Sekita, Y. et al. Role of retrotransposon-derived imprinted gene, Rtl1, in the feto-maternal interface of mouse placenta. *Nat Genet* 40, 243-8 (2008).

913. Esnault, C. et al. A placenta-specific receptor for the fusogenic, endogenous retrovirus-derived, human syncytin-2. *Proc Natl Acad Sci USA* 105, 17532-7 (2008).

914. Heidmann, O., Vernochet, C., Dupressoir, A. and Heidmann, T. Identification of an endogenous retroviral envelope gene with fusogenic activity and placenta-specific expression in the rabbit: a new "syncytin" in a third order of mammals. *Retrovirology* 6, 107 (2009).

915. Noorali, S. et al. Role of HERV-W syncytin-1 in placentation and maintenance of human pregnancy. *Appl Immunohistochem Mol Morphol* 17, 319-28 (2009).

916. Blackburn, D.G. Saltationist and punctuated equilibrium models for the evolution of viviparity and placentation. *J Theoretical Biol* 174, 199-216 (1995).

917. Beraldi, R., Pittoggi, C., Sciamanna, I., Mattei, E. and Spadafora, C. Expression of LINE-1 retroposons is essential for murine preimplantation development. *Mol Reprod Dev* 73, 279-87 (2006).

918. Perrin, D. et al. Specific hypermethylation of LINE-1 elements during abnormal overgrowth and differentiation of human placenta. *Oncogene* 26, 2518-24 (2007).

919. Kaneko-Ishino, T. and Ishino, F. Retrotransposon silencing by DNA methylation contributed to the evolution of placentation and genomic imprinting in mammals. *Dev Growth Differ* 52, 533-43 (2010).

920. Wessler, S.R., Baran, G. and Varagona, M. The maize transposable element Ds is spliced from RNA. *Science* 237, 916-8 (1987).

921. Wessler, S.R. The maize transposable Ds1 element is alternatively spliced from exon sequences. *Mol Cell Biol* 11, 6192-6 (1991).

922. Zabala, G. and Vodkin, L. Novel exon combinations generated by alternative splicing of gene fragments mobilized by a CACTA transposon in Glycine max. *BMC Plant Biol* 7, 38 (2007).

923. Varagona, M.J., Purugganan, M. and Wessler, S.R. Alternative splicing induced by insertion of retrotransposons into the maize waxy gene. *Plant Cell* 4, 811-20 (1992).

924. Gal-Mark, N., Schwartz, S. and Ast, G. Alternative splicing of Alu exons—two arms are better than one. *Nucleic Acids Res* 36, 2012-23 (2008).

925. Kreahling, J. and Graveley, B.R. The origins and implications of Aluternative splicing. *Trends Genet* 20, 1-4 (2004).

926. Han, J.S., Szak, S.T. and Boeke, J.D. Transcriptional disruption by the L1 retrotransposon and implications for mammalian transcriptomes. *Nature* 429, 268-74 (2004).

927. Han, J.S. and Boeke, J.D. LINE-1 retrotransposons: modulators of quantity and quality of mammalian gene expression? *Bioessays* 27, 775-84 (2005).

928. Ustyugova, S.V., Lebedev, Y.B. and Sverdlov, E.D. Long L1 insertions in human gene introns specifically reduce the content of corresponding primary transcripts. *Genetica* 128, 261-72 (2006).

929. Comfort, N.C. "The real point is control": The reception of Barbara McClintock's controlling elements. *J Hist Biol* 32, 133-6 (1999).

930. Shapiro, J.A. Mutations caused by the insertion of genetic material into the galactose operon of Escherichia coli. *J Mol Biol* 40, 93-105 (1969).

931. De Crombrugghe B, A.S., Gottesman M, Pastan I. Effect of Rho on transcription of bacterial operons. *Nature New Biology* 241, 260-4 (1973).

932. Pilacinski, W. et al. Insertion sequence IS2 associated with int-constitutive mutants of bacteriophage lambda. *Gene* 2, 61-74 (1977).

933. Nusse, R. and Varmus, H.E. Many tumors induced by the mouse mammary tumor virus contain a provirus integrated in the same region of the host genome. *Cell* 31, 99-109 (1982).

934. Nusse, R. Insertional mutagenesis in mouse mammary tumorigenesis. *Curr Top Microbiol Immunol* 171, 43-65 (1991).

935. Tsichlis, P.N. et al. Activation of multiple genes by provirus integration in the Mlvi-4 locus in T-cell lymphomas induced by Moloney murine leukemia virus. *J Virol* 64, 2236-44 (1990).

936. Marino-Ramirez, L. and Jordan, I.K. Transposable element derived DNaseI-hypersensitive sites in the human genome. *Biol Direct* 1, 20 (2006).

937. Romanish, M.T., Lock, W.M., van de Lagemaat, L.N., Dunn, C.A. and Mager, D.L. Repeated recruitment of LTR retrotransposons as promoters by the anti-apoptotic locus NAIP during mammalian evolution. *PLoS Genet* 3, e10 (2007).

938. Qureshi, I.A. and Mehler, M.F. Regulation of non-coding RNA networks in the nervous system—what's the REST of the story? *Neurosci Lett* 466, 73-80 (2009).

939. Johnson, R. et al. Evolution of the vertebrate gene regulatory network controlled by the transcriptional repressor REST. *Mol Biol Evol* 26, 1491-507 (2009).

940. Johnson, R. et al. Identification of the REST regulon reveals extensive transposable element-mediated binding site duplication. *Nucleic Acids Res* 34, 3862-77 (2006).

941. Peaston, A.E. et al. Retrotransposons regulate host genes in mouse oocytes and preimplantation embryos. *Dev Cell* 7, 597-606 (2004).

942. Wang, J., Bowen, N.J., Marino-Ramirez, L. and Jordan, I.K. A c-Myc regulatory subnetwork from human transposable element sequences. *Mol Biosyst* 5, 1831-9 (2009).

943. Slotkin, R.K. and Martienssen, R. Transposable elements and the epigenetic regulation of the genome. *Nat Rev Genet* 8, 272-85 (2007).

944. Fontanillas, P., Hartl, D.L. and Reuter, M. Genome organization and gene expression shape the transposable element distribution in the Drosophila melanogaster euchromatin. *PLoS Genet* 3, e210 (2007).

945. Huda, A. and Jordan, I.K. Epigenetic regulation of Mammalian genomes by transposable elements. *Annu NY Acad Sci* 1178, 276-84 (2009).

946. Gao, X., Hou, Y., Ebina, H., Levin, H.L. and Voytas, D.F. Chromodomains direct integration of retrotransposons to heterochromatin. *Genome Res* 18, 359-69 (2008).

947. Bennetzen, J.L. The structure and evolution of angiosperm nuclear genomes. *Curr Opin Plant Biol* 1, 103-8 (1998).

948. Nagaki, K. et al. Structure, divergence, and distribution of the CRR centromeric retrotransposon family in rice. *Mol Biol Evol* 22, 845-55 (2005).

949. Kanizay, L. and Dawe, R.K. Centromeres: long intergenic spaces with adaptive features. *Funct Integr Genomics* 9, 287-92 (2009).

950. Wolfgruber, T.K. et al. Maize Centromere Structure and Evolution: Sequence Analysis of Centromeres 2 and 5 Reveals Dynamic Loci Shaped Primarily by Retrotransposons. *PLoS Genet* 5, e1000743 (2009).

951. Walter, J., Hutter, B., Khare, T. and Paulsen, M. Repetitive elements in imprinted genes. *Cytogenet Genome Res* 113, 109-15 (2006).

952. Suzuki, S. et al. Retrotransposon silencing by DNA methylation can drive mammalian genomic imprinting. *PLoS Genet* 3, e55 (2007).

953. Fujimoto, R. et al. Evolution and control of imprinted FWA genes in the genus Arabidopsis. *PLoS Genet* 4, e1000048 (2008).

954. Pask, A.J. et al. Analysis of the platypus genome suggests a transposon origin for mammalian imprinting. *Genome Biol* 10, R1 (2009).

955. Gehring, M., Bubb, K.L. and Henikoff, S. Extensive demethylation of repetitive elements during seed development underlies gene imprinting. *Science* 324, 1447-51 (2009).

956. Amaral, P.P., Dinger, M.E., Mercer, T.R. and Mattick, J.S. The eukaryotic genome as an RNA machine. *Science* 319, 1787-9 (2008).

957. Glinsky, G.V. Phenotype-defining functions of multiple non-coding RNA pathways. *Cell Cycle* 7, 1630-9 (2008).

958. Zuckerkandl, E. and Cavalli, G. Combinatorial epigenetics, "junk DNA," and the evolution of complex organisms. *Gene* 390, 232-42 (2007).

959. Mattick, J.S. The genetic signatures of noncoding RNAs. *PLoS Genet* 5, e1000459 (2009).

960. Piriyapongsa, J. and Jordan, I.K. Dual coding of siRNAs and miRNAs by plant transposable elements. *RNA* 14, 814-21 (2008).

961. Kuang, H. et al. Identification of miniature inverted-repeat transposable elements (MITEs) and biogenesis of their siRNAs in the Solanaceae: new functional implications for MITEs. *Genome Res* 19, 42-56 (2009).

962. Hanada, K. et al. The functional role of pack-MULEs in rice inferred from purifying selection and expression profile. *Plant Cell* 21, 25-38 (2009).

963. Obbard, D.J., Gordon, K.H., Buck, A.H. and Jiggins, F.M. The evolution of RNAi as a defence against viruses and transposable elements. *Philos Trans R Soc Lond B Biol Sci* 364, 99-115 (2009).

964. Smalheiser, N.R. and Torvik, V.I. Mammalian microRNAs derived from genomic repeats. *Trends Genet* 21, 322-6 (2005).

965. Mariner, P.D. et al. Human Alu RNA is a modular transacting repressor of mRNA transcription during heat shock. *Mol Cell* 29, 499-509 (2008).

966. Yakovchuk, P., Goodrich, J.A. and Kugel, J.F. B2 RNA and Alu RNA repress transcription by disrupting contacts between RNA polymerase II and promoter DNA within assembled complexes. *Proc Natl Acad Sci USA* 106, 5569-74 (2009).

967. Lehnert, S. et al. Evidence for co-evolution between human microRNAs and Alu-repeats. *PLoS One* 4, e4456 (2009).

968. Devor, E.J., Peek, A.S., Lanier, W. and Samollow, P.B. Marsupial-specific microRNAs evolved from marsupial-specific transposable elements. *Gene* 448, 187-91 (2009).

969. White, M.J. Chromosomes of the vertebrates. *Evolution* 3, 379-81 (1949).

970. Carson, H.L., Clayton, F.E. and Stalker, H.D. Karyotypic stability and speciation in Hawaiian Drosophila. *Proc Natl Acad Sci USA* 57, 1280-5 (1967).

971. Stalker, H.D. Intergroup phylogenies in Drosophila as determined by comparisons of salivary banding patterns. *Genetics* 70, 457-74 (1972).

972. Nakatani, Y., Takeda, H., Kohara, Y. and Morishita, S. Reconstruction of the vertebrate ancestral genome reveals dynamic genome reorganization in early vertebrates. *Genome Res* 17, 1254-65 (2007).

973. Schubert, I. Chromosome evolution. *Curr Opin Plant Biol* 10, 109-15 (2007).

974. Kemkemer, C. et al. Gene synteny comparisons between different vertebrates provide new insights into breakage and fusion events during mammalian karyotype evolution. *BMC Evol Biol* 9, 84 (2009).

975. Kikuta, H. et al. Genomic regulatory blocks encompass multiple neighboring genes and maintain conserved synteny in vertebrates. *Genome Res* 17, 545-55 (2007).

976. Olinski, R.P., Lundin, L.G. and Hallbook, F. Conserved synteny between the Ciona genome and human paralogons identifies large duplication events in the molecular evolution of the insulin-relaxin gene family. *Mol Biol Evol* 23, 10-22 (2006).

977. Tang, H. Synteny and collinearity in plant genomes. *Science* 320, 486-488 (2008).

978. Waterston, R.H. et al. Initial sequencing and comparative analysis of the mouse genome. *Nature* 420, 520-62 (2002).

979. Samonte, R.V. and Eichler, E.E. Segmental duplications and the evolution of the primate genome. *Nat Rev Genet* 3, 65-72 (2002).

980. Koszul, R. and Fischer, G. A prominent role for segmental duplications in modeling eukaryotic genomes. *C R Biol* 332, 254-66 (2009).

981. Coulibaly, M.B. et al. Segmental duplication implicated in the genesis of inversion 2Rj of Anopheles gambiae. *PLoS One* 2, e849 (2007).

982. Stankiewicz, P., Shaw, C.J., Withers, M., Inoue, K. and Lupski, J.R. Serial segmental duplications during primate evolution result in complex human genome architecture. *Genome Res* 14, 2209-20 (2004).

983. Wolfe, K.H. Yesterday's polyploids and the mystery of diploidization. *Nat Rev Genet* 2, 333-41 (2001).

984. Ma, X.F. and Gustafson, J.P. Genome evolution of allopolyploids: a process of cytological and genetic diploidization. *Cytogenet Genome Res* 109, 236-49 (2005).

985. Tate, J.A., Joshi, P., Soltis, K.A., Soltis, P.S. and Soltis, D.E. On the road to diploidization? Homoeolog loss in independently formed populations of the allopolyploid Tragopogon miscellus (Asteraceae). *BMC Plant Biol* 9, 80 (2009).

986. Carbone, L. et al. Evolutionary breakpoints in the gibbon suggest association between cytosine methylation and karyotype evolution. *PLoS Genet* 5, e1000538 (2009).

987. Lemaitre, C. et al. Analysis of fine-scale mammalian evolutionary breakpoints provides new insight into their relation to genome organisation. *BMC Genomics* 10, 335 (2009).

988. Sankoff, D. The where and wherefore of evolutionary breakpoints. *J Biol Chem* 8, 66 (2009).

989. Larkin, D.M. et al. Breakpoint regions and homologous synteny blocks in chromosomes have different evolutionary histories. *Genome Res* 19, 770-7 (2009).

990. Glover, T.W. Common fragile sites. *Cancer Lett* 232, 4-12 (2006).

991. Arlt, M.F., Durkin, S.G., Ragland, R.L. and Glover, T.W. Common fragile sites as targets for chromosome rearrangements. *DNA Repair (Amst)* 5, 1126-35 (2006).

992. Ruiz-Herrera, A. and Robinson, T.J. Chromosomal instability in Afrotheria: fragile sites, evolutionary breakpoints and phylogenetic inference from genome sequence assemblies. *BMC Evol Biol* 7, 199 (2007).

993. Durkin, S.G. and Glover, T.W. Chromosome fragile sites. *Annu Rev Genet* 41, 169-92 (2007).

994. von Grotthuss, M., Ashburner, M. and Ranz, J.M. Fragile regions and not functional constraints predominate in shaping gene organization in the genus Drosophila. *Genome Res* 20, 1084-96 (2010).

995. Rachidi, N., Barre, P. and Blondin, B. Multiple Ty-mediated chromosomal translocations lead to karyotype changes in a wine strain of Saccharomyces cerevisiae. *Mol Gen Genet* 261, 841-50 (1999).

996. Longo, M.S., Carone, D.M., Green, E.D., O'Neill, M.J. and O'Neill, R.J. Distinct retroelement classes define evolutionary breakpoints demarcating sites of evolutionary novelty. *BMC Genomics* 10, 334 (2009).

997. Kehrer-Sawatzki, H. and Cooper, D.N. Molecular mechanisms of chromosomal rearrangement during primate evolution. *Chromosome Res* 16, 41-56 (2008).

998. Caceres, M., Ranz, J.M., Barbadilla, A., Long, M. and Ruiz, A. Generation of a widespread Drosophila inversion by a transposable element. *Science* 285, 415-8 (1999).

999. Evgen'ev, M.B. et al. Mobile elements and chromosomal evolution in the virilis group of Drosophila. *Proc Natl Acad Sci USA* 97, 11337-42 (2000).

1000. Caceres, M., Puig, M. and Ruiz, A. Molecular characterization of two natural hotspots in the Drosophila buzzatii genome induced by transposon insertions. *Genome Res* 11, 1353-64 (2001).

1001. Lyttle, T.W. and Haymer, D.S. The role of the transposable element hobo in the origin of endemic inversions in wild populations of Drosophila melanogaster. *Genetica* 86, 113-26 (1992).

1002. Delprat, A., Negre, B., Puig, M. and Ruiz, A. The transposon Galileo generates natural chromosomal inversions in Drosophila by ectopic recombination. *PLoS One* 4, e7883 (2009).

1003. Girirajan, S. et al. Sequencing human-gibbon breakpoints of synteny reveals mosaic new insertions at rearrangement sites. *Genome Res* 19, 178-90 (2009).

1004. Yunis, J.J. Multiple recurrent genomic rearrangements and fragile sites in human cancer. *Somat Cell Mol Genet* 13, 397-403 (1987).

1005. Le Beau, M.M. Chromosomal fragile sites and cancer-specific rearrangements. *Blood* 67, 849-58 (1986).

1006. Pichiorri, F. et al. Molecular parameters of genome instability: roles of fragile genes at common fragile sites. *J Cell Biochem* 104, 1525-33 (2008).

1007. Zhang, J. et al. Alternative Ac/Ds transposition induces major chromosomal rearrangements in maize. *Genes Dev* 23, 755-65 (2009).

1008. Cuvier, G. *Le règne animal distribué d'après son organisation, pour servir de base à l'histoire naturelle des animaux et d'introduction à l'anatomie comparée*, (Deterville, Paris, 1817).

1009. Cuvier, G., M'Murtrie, H. (transl.). *The Animal Kingdom, Arranged in Conformity with its Organization*, (G. and C. and H. Carvill, New York, 1832).

1010. Darwin, E. *The Botanic Garden, Part II, The Loves of the Plants*, (J. Johnson, London, 1789).

1011. Darwin, E. *Zoonomia; or, The Laws of Organic Life. Part I* (J. Johnson, London, 1794).

1012. Darwin, E. *Zoonomia; or, The Laws of Organic Life. Part II* (London, 1796).

1013. Darwin, E. *The Temple of Nature; or, The Origin of Society* (J. Johnson, 1806–1807).

1014. Carroll, S.B., Prud'homme, B. and Gompel, N. Regulating Evolution. *Sci Am* 298, 60-67 (2008).

1015. Muller, G.B. and Newman, S.A. The innovation triad: an EvoDevo agenda. *J Exp Zool B Mol Dev Evol* 304, 487-503 (2005).

1016. De Robertis, E.M. Evo-devo: variations on ancestral themes. *Cell* 132, 185-95 (2008).

1017. Carroll, S.B. Evo-devo and an expanding evolutionary synthesis: a genetic theory of morphological evolution. *Cell* 134, 25-36 (2008).

1018. Kuratani, S. Modularity, comparative embryology and evo-devo: developmental dissection of evolving body plans. *Dev Biol* 332, 61-9 (2009).

1019. Gehring, W.J. The Molecular Basis of Development. *Sci Am* 253, 152-162 (1985).

1020. De Robertis, E.M., Oliver, G. and Wright, C.V. Homeobox genes and the vertebrate body plan. *Sci Am* 263, 46-52 (1990).

1021. McGinnis, W. and Kuziora, M. The Molecular Architects of Body Design. *Sci Am* 270, 58-66 (1994).

1022. Holland, P.W. and Garcia-Fernandez, J. Hox genes and chordate evolution. *Dev Biol* 173, 382-395 (1996).

1023. Lemons, D. and McGinnis, W. Genomic evolution of Hox gene clusters. *Science* 313, 1918-22 (2006).

1024. Ryan, J.F. and Baxevanis, A.D. Hox, Wnt, and the evolution of the primary body axis: insights from the early-divergent phyla. *Biol Direct* 2, 37 (2007).

1025. Kuraku, S. and Meyer, A. The evolution and maintenance of Hox gene clusters in vertebrates and the teleost-specific genome duplication. *Int J Dev Biol* 53, 765-73 (2009).

1026. Maeda, R.K. and Karch, F. The bithorax complex of Drosophila an exceptional Hox cluster. *Curr Top Dev Biol* 88, 1-33 (2009).

1027. Duboule, D. and Dolle, P. The structural and functional organization of the murine HOX gene family resembles that of Drosophila homeotic genes. *Embo J* 8, 1497-505 (1989).

1028. Izpisua-Belmonte, J.C., Falkenstein, H., Dolle, P., Renucci, A. and Duboule, D. Murine genes related to the Drosophila AbdB homeotic genes are sequentially expressed during development of the posterior part of the body. *Embo J* 10, 2279-89 (1991).

1029. McGinnis, N., Kuziora, M.A. and McGinnis, W. Human Hox-4.2 and Drosophila deformed encode similar regulatory specificities in Drosophila embryos and larvae. *Cell* 63, 969-76 (1990).

1030. Malicki, J., Schughart, K. and McGinnis, W. Mouse Hox-2.2 specifies thoracic segmental identity in Drosophila embryos and larvae. *Cell* 63, 961-7 (1990).

1031. Li, E. and Davidson, E.H. Building developmental gene regulatory networks. *Birth Defects Res C Embryo Today* 87, 123-30 (2009).

1032. Ho, M.C. et al. Functional evolution of cis-regulatory modules at a homeotic gene in Drosophila. *PLoS Genet* 5, e1000709 (2009).

1033. Garcia-Fernandez, J. The genesis and evolution of homeobox gene clusters. *Nat Rev Genet* 6, 881-92 (2005).

1034. Valentine, J.W., Jablonski, D. and Erwin, D.H. Fossils, molecules and embryos: new perspectives on the Cambrian explosion. *Development* 126, 851-9 (1999).

1035. Valentine, J.W. and Jablonski, D. Morphological and developmental macroevolution: a paleontological perspective. *Int J Dev Biol* 47, 517-22 (2003).

1036. Finnerty, J.R. The origins of axial patterning in the metazoa: how old is bilateral symmetry? *Int J Dev Biol* 47, 523-9 (2003).

1037. Finnerty, J.R., Pang, K., Burton, P., Paulson, D. and Martindale, M.Q. Origins of bilateral symmetry: Hox and dpp expression in a sea anemone. *Science* 304, 1335-7 (2004).

1038. Garcia-Fernandez, J. Hox, ParaHox, ProtoHox: facts and guesses. *Heredity* 94, 145-52 (2005).

1039. Manuel, M. Early evolution of symmetry and polarity in metazoan body plans. *C R Biol* 332, 184-209 (2009).

1040. Thomas-Chollier, M., Ledent, V., Leyns, L. and Vervoort, M. A non-tree-based comprehensive study of metazoan Hox and ParaHox genes prompts new insights into their origin and evolution. *BMC Evol Biol* 10, 73 (2010).

1041. Chiori, R. et al. Are Hox genes ancestrally involved in axial patterning? Evidence from the hydrozoan Clytia hemisphaerica (Cnidaria). *PLoS One* 4, e4231 (2009).

1042. Knoll, A.H., Javaux, E.J., Hewitt, D. and Cohen, P. Eukaryotic organisms in Proterozoic oceans. *Philos Trans R Soc Lond B Biol Sci* 361, 1023-38 (2006).

1043. Xiao, S. and Laflamme, M. On the eve of animal radiation: phylogeny, ecology and evolution of the Ediacara biota. *Trends Ecol Evol* 24, 31-40 (2009).

1044. Peterson, K.J., Cotton, J.A., Gehling, J.G. and Pisani, D. The Ediacaran emergence of bilaterians: congruence between the genetic and the geological fossil records. *Philos Trans R Soc Lond B Biol Sci* 363, 1435-43 (2008).

1045. Conway Morris, S. and Whittington, H.B. The Animals of the Burgess Shale. *Sci Am* 241, 122-135 (1979).

1046. Levinton, J.S. The Big Bang of Animal Evolution. *Sci Am* 267, 84-91 (1992).

1047. Freeman, G. A developmental basis for the Cambrian radiation. *Zoolog Sci* 24, 113-22 (2007).

1048. Budd, G.E. and Jensen, S. A critical reappraisal of the fossil record of the bilaterian phyla. *Biol Rev Camb Philos Soc* 75, 253-95 (2000).

1049. Conway Morris, S. Darwin's dilemma: the realities of the Cambrian "explosion." *Philos Trans R Soc Lond B Biol Sci* 361, 1069-83 (2006).

1050. Brysse, K. From weird wonders to stem lineages: the second reclassification of the Burgess Shale fauna. *Stud Hist Philos Biol Biomed Sci* 39, 298-313 (2008).

1051. Chen, J.Y. et al. Complex embryos displaying bilaterian characters from Precambrian Doushantuo phosphate deposits, Weng'an, Guizhou, China. *Proc Natl Acad Sci USA* 106, 19056-60 (2009).

1052. Wigglesworth, V.B. Metamorphosis, Polymorphism, Differentiation. *Sci Am* 200, 100-110 (1959).

1053. Hadorn, E. Transdetermination in Cells. *Sci Am* 219, 110-120 (1968).

1054. García-Bellido, A., Lawrence, P.A. and Morata, G. Compartments in Animal Development. *Sci Am* 241, 102-111 (1979).

1055. Nusslein-Volhard, C. Gradients that organize embryo development. *Sci Am* 275, 54-5; 58-61 (1996).

1056. Nusslein-Volhard, C. and Wieschaus, E. Mutations affecting segment number and polarity in Drosophila. *Nature* 287, 795-801 (1980).

1057. Gordon, R. and Jacobson, A.G. The Shaping of Tissues in Embryos. *Sci Am* 238, 106-113 (1978).

1058. Riddle, R.D. and Tabin, C.J. How Limbs Develop. *Sci Am* 280, 74-79 (1999).

1059. Duboule, D. and Wilkins, A.S. The evolution of "bricolage," *Trends Genet* 14, 54-9 (1998).

1060. Wilkins, A.S. *The Evolution of Developmental Pathways*, (Sinauer, 2002).

1061. Wilkins, A.S. Recasting developmental evolution in terms of genetic pathway and network evolution ... and the implications for comparative biology. *Brain Res Bull* 66, 495-509 (2005).

1062. Wilkins, A.S. Between "design" and "bricolage": genetic networks, levels of selection, and adaptive evolution. *Proc Natl Acad Sci USA* 104 Suppl 1, 8590-6 (2007).

1063. Albert, M. and Peters, A.H. Genetic and epigenetic control of early mouse development. *Curr Opin Genet Dev* 19, 113-21 (2009).

1064. Mattick, J.S. A new paradigm for developmental biology. *J Exp Biol* 210, 1526-47 (2007).

1065. Vasanthi, D. and Mishra, R.K. Epigenetic regulation of genes during development: a conserved theme from flies to mammals. *J Genet Genomics* 35, 413-29 (2008).

1066. Vavouri, T., Walter, K., Gilks, W.R., Lehner, B. and Elgar, G. Parallel evolution of conserved non-coding elements that target a common set of developmental regulatory genes from worms to humans. *Genome Biol* 8, R15 (2007).

1067. Borok, M.J., Tran, D.A., Ho, M.C. and Drewell, R.A. Dissecting the regulatory switches of development: lessons from enhancer evolution in Drosophila. *Development* 137, 5-13 (2010).

1068. Gehring, W.J., Kloter, U. and Suga, H. Evolution of the Hox gene complex from an evolutionary ground state. *Curr Top Dev Biol* 88, 35-61 (2009).

1069. Mungpakdee, S. et al. Differential evolution of the 13 Atlantic salmon Hox clusters. *Mol Biol Evol* 25, 1333-43 (2008).

1070. Spring, J. Genome duplication strikes back. *Nat Genet* 31, 128-9 (2002).

1071. Ohno, S. *Evolution by Gene Duplication* (George Allen and Unwin, London, 1970).

1072. Volff, J.N. and Schartl, M. Evolution of signal transduction by gene and genome duplication in fish. *J Struct Funct Genomics* 3, 139-50 (2003).

1073. Dehal, P. and Boore, J.L. Two rounds of whole genome duplication in the ancestral vertebrate. *PLoS Biol* 3, e314 (2005).

1074. Kawasaki, K., Buchanan, A.V. and Weiss, K.M. Gene duplication and the evolution of vertebrate skeletal mineralization. *Cells Tissues Organs* 186, 7-24 (2007).

1075. Hulse, J.H. and Spurgeon, D. Triticale. *Sci Am* 231, 72-80 (1974).

1076. Wilson, A. Wheat and rye hybrids. *Edinburgh Botanical Society Transactions* 12, 286-288 (1876).

1077. Meister, G. Natural hybridization of wheat and rye in Russia. *J Hered* 12, 467-470 (1921).

1078. Ma, X.F. and Gustafson, J.P. Allopolyploidization-accommodated genomic sequence changes in triticale. *Ann Bot (Lond)* 101, 825-32 (2008).

1079. Anderson, E., Stebbins, G.L., Jr. Hybridization as an evolutionary stimulus. *Evolution* 8, 378-388 (1954).

1080. Arnold, M.L. Transfer and origin of adaptations through natural hybridization: were Anderson and Stebbins right? *Plant Cell* 16, 562-70 (2004).

1081. Crow, J.F. Plant breeding giants. Burbank, the artist; Vavilov, the scientist. *Genetics* 158, 1391-5 (2001).

1082. Eakin, G.S. and Behringer, R.R. Tetraploid development in the mouse. *Dev Dyn* 228, 751-66 (2003).

1083. Beaulieu, J., Jean, M. and Belzile, F. The allotetraploid Arabidopsis thaliana-Arabidopsis lyrata subsp. petraea as an alternative model system for the study of polyploidy in plants. *Mol Genet Genomics* 281, 421-35 (2009).

1084. Mallet, J., Beltran, M., Neukirchen, W. and Linares, M. Natural hybridization in heliconiine butterflies: the species boundary as a continuum. *BMC Evol Biol* 7, 28 (2007).

1085. Mallet, J. Hybrid speciation. *Nature* 446, 279-83 (2007).

1086. Dasmahapatra, K.K., Silva-Vasquez, A., Chung, J.W. and Mallet, J. Genetic analysis of a wild-caught hybrid between non-sister Heliconius butterfly species. *Biol Lett* 3, 660-3 (2007).

1087. Salazar, C. et al. Genetic evidence for hybrid trait speciation in heliconius butterflies. *PLoS Genet* 6, e1000930 (2010).

1088. Darwin, F., Seward, AC, eds. *More Letters of Charles Darwin* (John Murray, London, 1903).

1089. Cui, L. Widespread genome duplications throughout the history of flowering plants. *Genome Res* 16, 738-749 (2006).

1090. Doyle, J.J. et al. Evolutionary genetics of genome merger and doubling in plants. *Annu Rev Genet* 42, 443-61 (2008).

1091. Lim, K.Y. et al. Rapid chromosome evolution in recently formed polyploids in Tragopogon (Asteraceae). *PLoS One* 3, e3353 (2008).

1092. Soltis, D.E. Polyploidy and angiosperm diversification. *Am J Bot* 96, 336-348 (2009).

1093. Wolfe, K.H. and Shields, D.C. Molecular evidence for an ancient duplication of the entire yeast genome. *Nature* 387, 708-713 (1997).

1094. Musso, G., Zhang, Z. and Emili, A. Retention of protein complex membership by ancient duplicated gene products in budding yeast. *Trends Genet* 23, 266-9 (2007).

1095. Conant, G.C. and Wolfe, K.H. Increased glycolytic flux as an outcome of whole-genome duplication in yeast. *Mol Syst Biol* 3, 129 (2007).

1096. Aury, J.M. et al. Global trends of whole-genome duplications revealed by the ciliate Paramecium tetraurelia. *Nature* 444, 171-178 (2006).

1097. Veron, A.S., Kaufmann, K. and Bornberg-Bauer, E. Evidence of interaction network evolution by whole-genome duplications: a case study in MADS-box proteins. *Mol Biol Evol* 24, 670-8 (2007).

1098. Gillis, W.Q., John, J.S., Bowerman, B. and Schneider, S.Q. Whole genome duplications and expansion of the vertebrate GATA transcription factor gene family. *BMC Evol Biol* 9, 207 (2009).

1099. Wilkins, A.S. Genetic networks as transmitting and amplifying devices for natural genetic tinkering. Novartis Found Symp 284, 71-86; discussion 86-9, 110-5 (2007).

1100. Kassahn, K.S., Dang, V.T., Wilkins, S.J., Perkins, A.C. and Ragan, M.A. Evolution of gene function and regulatory control after whole-genome duplication: comparative analyses in vertebrates. *Genome Res* 19, 1404-18 (2009).

1101. Bray, D. Intracellular signalling as a parallel distributed process. *J Theor Biol* 143, 215-31 (1990).

1102. Morelli, M.J., Ten Wolde, P.R. and Allen, R.J. DNA looping provides stability and robustness to the bacteriophage lambda switch. *Proc Natl Acad Sci USA* 106, 8101-6 (2009).

1103. Jacob, F. Evolution and tinkering. *Science* 196, 1161-6 (1977).

1104. Bubanovic, I., Najman, S. and Andjelkovic, Z. Origin and evolution of viruses: escaped DNA/RNA sequences as evolutionary accelerators and natural biological weapons. *Med Hypotheses* 65, 868-72 (2005).

1105. Claverie, J.M. Viruses take center stage in cellular evolution. *Genome Biol* 7, 110 (2006).

1106. Forterre, P. The origin of viruses and their possible roles in major evolutionary transitions. *Virus Res* 117, 5-16 (2006).

1107. Forterre, P. and Gadelle, D. Phylogenomics of DNA topoisomerases: their origin and putative roles in the emergence of modern organisms. *Nucleic Acids Res* 37, 679-92 (2009).

1108. Koonin, E.V., Senkevich, T.G. and Dolja, V.V. The ancient Virus World and evolution of cells. *Biol Direct* 1, 29 (2006).

1109. Koonin, E.V. Temporal order of evolution of DNA replication systems inferred by comparison of cellular and viral DNA polymerases. *Biol Direct* 1, 39 (2006).

1110. Delaroque, N. and Boland, W. The genome of the brown alga Ectocarpus siliculosus contains a series of viral DNA pieces, suggesting an ancient association with large dsDNA viruses. *BMC Evol Biol* 8, 110 (2008).

1111. Bell, P.J. The viral eukaryogenesis hypothesis: a key role for viruses in the emergence of eukaryotes from a prokaryotic world environment. *Annu NY Acad Sci* 1178, 91-105 (2009).

1112. Keller, J. et al. A protein encoded by a new family of mobile elements from Euryarchaea exhibits three domains with novel folds. *Protein Sci* 18, 850-5 (2009).

1113. Groves, D.I., Dunlop, J.S.R. and Buick, R. An Early Habitat of Life. *Sci Am* 245, 64-73 (1981).

1114. Allwood, A.C., Walter, M.R., Kamber, B.S., Marshall, C.P. and Burch, I.W. Stromatolite reef from the Early Archaean era of Australia. *Nature* 441, 714-8 (2006).

1115. Trewavas, A. Plant intelligence. *Naturwissenschaften* 92, 401-13 (2005).

1116. Brenner, E.D. et al. Plant neurobiology: an integrated view of plant signaling. *Trends Plant Sci* 11, 413-9 (2006).

1117. Barlow, P.W. Reflections on "plant neurobiology." *Biosystems* 92, 132-47 (2008).

1118. Cvrckova, F., Lipavska, H. and Zarsky, V. Plant intelligence: why, why not or where? *Plant Signal Behav* 4, 394-9 (2009).

1119. Perkins, T.J. and Swain, P.S. Strategies for cellular decision-making. *Mol Syst Biol* 5, 326 (2009).

1120. Darwin, C.R. *The Power of Movements in Plants* (John Murray, London, 1880).

1121. Baluska, F., Mancuso, S., Volkmann, D. and Barlow, P.W. The "root-brain" hypothesis of Charles and Francis Darwin: Revival after more than 125 years. *Plant Signal Behav* 4, 1121-7 (2009).

1122. Campbell, J.H., Lengyel, J.A. and Langridge, J. Evolution of a second gene for beta-galactosidase in Escherichia coli. *Proc Natl Acad Sci USA* 70, 1841-5 (1973).

1123. Hall, B.G. and Hauer, B. Acquisition of new metabolic activities by microbial populations. *Methods Enzymol* 224, 603-13 (1993).

1124. Hunt, P. et al. Experimental evolution, genetic analysis and genome re-sequencing reveal the mutation conferring artemisinin resistance in an isogenic lineage of malaria parasites. *BMC Genomics* 11, 499 (2010).

1125. Comai, L. et al. Phenotypic instability and rapid gene silencing in newly formed arabidopsis allotetraploids. *Plant Cell* 12, 1551-68 (2000).

1126. Pontes, O. et al. Chromosomal locus rearrangements are a rapid response to formation of the allotetraploid Arabidopsis suecica genome. *Proc Natl Acad Sci USA* 101, 18240-5 (2004).

1127. Madlung, A. et al. Genomic changes in synthetic Arabidopsis polyploids. *Plant J* 41, 221-30 (2005).

1128. Wang, J. et al. Genomewide nonadditive gene regulation in Arabidopsis allotetraploids. *Genetics* 172, 507-17 (2006).

1129. Driesch, H. *The science and philosophy of the organism: Gifford lectures delivered at Aberdeen university, 1907–1908 by Hans Driesch* (printed for the University, Aberdeen, 1908).

1130. Melzer, S.J. Vitalism and mechanism in biology and medicine. *Science* 19, 18-22 (1904).

1131. Lillie, R.S. The philosophy of biology: vitalism versus mechanism. *Science* 40, 840-846 (1914).

1132. Holliday, R. Physics and the origins of molecular biology. *J Genet Genomics* 85, 93-7 (2006).

1133. Wiener, N. *Cybernetics or Control and Communication in the Animal and the Machine*, (MIT Press, Cambridge, 1965).

1134. Hopfields, J.J. Neural networks and physical systems with emergent collective computational abilities. *Proc Natl Acad Sci USA* 79, 2554-2558 (1982).

1135. Alvarez, W. and Asaro, F. An Extraterrestrial Impact. *Sci Am* 263, 78-84 (1990).

1136. Alvarez, L.W., Alvarez, W., Asaro, F. and Michel, H.V. Extraterrestrial cause for the cretaceous-tertiary extinction. *Science* 208, 1095-108 (1980).

1137. Schulte, P. et al. The Chicxulub asteroid impact and mass extinction at the Cretaceous-Paleogene boundary. *Science* 327, 1214-8 (2010).

1138. Raup, D.M. and Sepkoski, J.J., Jr. Mass extinctions in the marine fossil record. *Science* 215, 1501-3 (1982).

1139. Raup, D.M. and Sepkoski, J.J., Jr. Periodic extinction of families and genera. *Science* 231, 833-6 (1986).

1140. Raup, D.M. and Boyajian, G.E. Patterns of generic extinction in the fossil record. *Paleobiology* 14, 109-25 (1988).

1141. Rampino, M.R., Haggerty, B.M. and Pagano, T.C. A unified theory of impact crises and mass extinctions: quantitative tests. *Annu NY Acad Sci* 822, 403-31 (1997).

1142. Sengor, A.M., Atayman, S. and Ozeren, S. A scale of greatness and causal classification of mass extinctions: implications for mechanisms. *Proc Natl Acad Sci USA* 105, 13736-40 (2008).

1143. Kirchner, J.W. and Weil, A. Correlations in fossil extinction and origination rates through geological time. *Proc Biol Sci* 267, 1301-9 (2000).

1144. Pave, A., Herve, J.C. and Schmidt-Laine, C. Mass extinctions, biodiversity explosions and ecological niches. *C R Biol* 325, 755-65 (2002).

1145. Alroy, J. Colloquium paper: dynamics of origination and extinction in the marine fossil record. *Proc Natl Acad Sci USA* 105 Suppl 1, 11536-42 (2008).

1146. Foote, M. and Raup, D.M. Fossil preservation and the stratigraphic ranges of taxa. *Paleobiology* 22, 121-40 (1996).

1147. Foote, M. and Sepkoski, J.J., Jr. Absolute measures of the completeness of the fossil record. *Nature* 398, 415-7 (1999).

1148. Payne, J.L. et al. Two-phase increase in the maximum size of life over 3.5 billion years reflects biological innovation and environmental opportunity. *Proc Natl Acad Sci USA* 106, 24-7 (2009).

1149. Mayhew, P.J., Jenkins, G.B. and Benton, T.G. A long-term association between global temperature and biodiversity, origination and extinction in the fossil record. *Proc Biol Sci* 275, 47-53 (2008).

1150. Cloud, P. The Biosphere. *Sci Am* 249, 176-189 (1983).

1151. Gould, S.J. The Evolution of Life on Earth. *Sci Am* 290, 92-100 (2004).

1152. Gould, S.J. Punctuated Equilibrium and the Fossil Record. *Science* 219, 439-440 (1983).

1153. Gould, S.J. and Eldredge, N. Punctuated equilibrium comes of age. *Nature* 366, 223-7 (1993).

1154. Zeh, D.W., Zeh, J.A. and Ishida, Y. Transposable elements and an epigenetic basis for punctuated equilibria. *Bioessays* 31, 715-26 (2009).

1155. David, L.A. and Alm, E.J. Rapid evolutionary innovation during an Archaean genetic expansion. *Nature* 469, 93-96 (2011).

1156. Holland, J.H. Genetic Algorithms. *Sci Am* 267, 66-72 (1992).

1157. Foster, J.A. Evolutionary computation. *Nat Rev Genet* 2, 428-36 (2001).

1158. Francois, P. and Hakim, V. Design of genetic networks with specified functions by evolution in silico. *Proc Natl Acad Sci USA* 101, 580-5 (2004).

1159. Rodrigo, G., Carrera, J. and Elena, S.F. Network design meets in silico evolutionary biology. *Biochimie* 92, 746-52 (2010).

1160. Muller-Hill, B. Towards a linguistics of DNA and protein. *Hist Philos Life Sci* 21, 53-63 (1999).

1161. Newman, S.A. and Bhat, R. Dynamical patterning modules: a "pattern language" for development and evolution of multicellular form. *Int J Dev Biol* 53, 693-705 (2009).

1162. Pinker, S. *The Language Instinct: The New Science of Language and Mind*, (Penguin, 1994).

Index

A

Ac-CoA, 149
accuracy of DNA replication, 12
acetyl group, 149
Acidithiobacillus ferrooxidans, 58
activated B cells, 67
adaptive immunity, 66-69, 149
aerobic metabolism, 141
Agrobacterium tumefaciens, 45
algae, symbiogenesis in, 103-104
alletetaploid, 149
allolactose, 10
allosteric transitions, 11
alternative splicing, 109-110
Alvarez, Luis, 139
Alvarez, Walter, 139
Ames, Bruce, 18
Ames test, 18
Anabaena, 58
Anaplasma marginale, 59
angiosperm, 149
animals
    genomic responses to changes in
      ploidy and interspecific
      hybridization, 75-78
    interspecific hybridization, 121-122
anneal, 149
annotation, 39, 149
anterior-posterior (AP), 150
antibiotic resistance, 90-95
antibodies
    antibody production, 66-69
    definition of, 150
antigenic variation, 56-59, 150

antigens, 150
AP (anterior-posterior), 150
apoptosis, 23-24, 150
apoptosis cascade, 150
Arabidopsis interspecific
    hybrids, 122
Arabidopsis thaliana, 40, 138
Archaea
    discovery of, 98-100, 125
    RNA-based defense against viruses
      and plasmids, 78
archaeabacteria, 99
ars (autonomous replication
    sequence), 36, 150
Asteraceae (Compositae), 75
atmospheric transformations,
    microbe-generated, 141
atypical sexual encounters, genomic
    responses to, 75
autonomous replication sequence
    (ars), 36, 150
autotetraploid, 150
Avery, Oswald, 44

B

B lymphocytes, 66-67
Bacillus subtilis, 58-60
bacteria. *See also specific bacteria*
    antibiotic resistance, 90-95
    antigenic variation, 56-59
    expression of HR (homologous
      recombination) proteins, 45
    phase variation, 56-59
    programmed cell death, 23

RNA-based defense against viruses
    and plasmids, 78
site-specific recombination
    explained, 46-47
    role in phase and antigenic
        variation, 56-59
    role in temperate
        bacteriophage life cycles,
        60-61
bacteriocytes, 101
bacteriophages, 61, 150
    site-specific recombination in, 60
    temperate bacteriophages, 170
Bacteroides fragilis, 59
base pair (bp), 151
Beadle, George, 28
Beijerinck, Martinus W., 100
Bifidobacterium, 55
Bilateria, 117, 140, 150
bioinformatics, 150
Blackburn, Elizabeth, 64
Bordetella bronchiseptica, 59
Bordetella bronchisepticum, 55
Borrelia burgdorferi, 59
Borrelia hermsi, 59
Borrelia, 57
boundary elements, 150
bp (base pair), 151
Brassica, 121
breakpoints, 113, 151

C

c-onc (cellular oncogene), 51
Caenorhabditis briggsae, 40
Caenorhabditis elegans, 40
Cambrian explosion, 117
cAMP (cyclic adenosine
    monophosphate), 10, 151
Campylobacter fetus
    antigenic variation, 59
    phase variation, 58
Canis domesticus, 40
Cartesian dualists, 151
Casadaban, Malcolm, 97
cassettes, 151
cataclysmic evolution, 121
CDS, 30
cell cycle checkpoints, 18-21, 137
cell death, programmed, 23-24, 164
cell heredity, 3-4

cell signaling
    role in programmed celled death,
        23-24
    S. cerevisiae mating pheromone
        response, 21-22
cellular differentiation, 32
cellular information acquisition,
    transmission, and processing, 7-8
    cell cycle checkpoints, 18-21, 137
    DNA repair and mutagenesis,
        14-16
        SOS response, 16-17
        Weigle mutagenesis, 15
    DNA replication proofreading
        accuracy of DNA
            replication, 12
        exonuclease proofreading, 12
        mismatch repair, 13-14
    E. coli sugar metabolism, 8-11
    intercellular signaling
        role in programmed celled
            death, 23-24
        S. cerevisiae *mating*
            pheromone response, 21-22
    intracellular molecular information
        transfer, 24-26
cellular oncogene (c-onc), 51
cellular regulation of natural genetic
    engineering, 69
    genomic responses to changes in
        ploidy and interspecific
        hybridization, 75-78
    interspecific hybridization, 121-122
    life history events that alter
        epigenome, 80-81
    RNA-based defense against viruses
        and plasmids, 78
    stimuli documented to activate
        natural genetic engineering, 70-74
centromeres, 151
Central Dogma of Molecular
    Biology, 24-25
centrioles, 106, 151
centromeres, 34, 37, 130, 133
centrosomes, 106, 151
chance, 1-2
checkpoints, 18-21, 137
chloramphenicol, 151
Chlorarachniophyta, 103
Chlorobium, 55

chloroplasts, 102
chromatin, 151
  chromatin domains, 111-112, 151
  chromatin formatting, 31-36, 130.
    See also epigenome
chromodomains, 111-112, 151
chromosomes, 151
  chromosome painting, 38, 152
  chromosome rearrangements,
    113-115
  chromosome replication, 60
cilia, 105, 152
ciliated protozoa, 62-65, 136
Ciona intestinalis, 40
cis-acting control regions, 132
cis-regulatory modules (CRMs), 11,
  31, 110-111, 152
Clark, A. John, 17
classes of DNA, 39-43
Clostridium difficile, 58
clustered regularly interspaced
  short palindromic repeats
  (CRISPR), 78-79, 152
coding sequence, 29-30
coding sequences (exons), 130
commensal, 152
competence, 152
completing DNA replication, 37
composite elements, 41
conjugation, 45
consortiums, 101
constant regions, 152
coordinated transcription, 152
Cretaceous period, 139
Crick, Francis, 24-25
CRISPR (clustered regularly
  interspaced short palindromic
  repeats), 78-79, 152
CRMs (*cis*-regulatory modules), 11,
  31, 110-111, 152
crp (cAMP receptor protein), 10, 152
cut-and-paste transposition, 152
Cuvier, Georges, 115
cyanobacteria, 153
cyclic adenosine monophosphate
  (cAMP), 10, 151
cytogenetics, origins of natural
  genetic engineering in, 2
cytotype, 77, 153

D

Darwin, Charles, 1, 122, 127
Darwinism, 127
death receptors, 24
Delbrück, Max, 146
denitrification, 141
DGRs (diversity-generating
  retroelements), 55, 153
  explained, 55
  role in phase and antigenic
    variation, 57
diauxy, 9
Dichelobacter nodosus, 59
dinosaurs, disappearance of, 139
diploidization, 114, 153
diploids, 153
dispersed repeats, 153
DIVAC (diversification
  activator), 153
diversification of evolutionary
  inventions, 133-134
diversity, 5
diversity-generating retroelements
  (DGRs), 55-57, 153
DNA classes, 39-43
DNA import and export systems,
  44-45
DNA regulation. *See* cellular
  regulation of natural genetic
  engineering
DNA repair and mutagenesis, 14-16
  SOS response, 16-17
  Weigle mutagenesis, 15
DNA replication
  autonomous replication sequences
    (ars), 36
  centromeres, 37
  completion, 37
  ensuring transmission at cell
    division, 37
  initiation, 36
  origin of replication complex
    (ORC), 36
  origin of replication (ori) sites, 36
  partition (par) sites, 37
  proofreading
    *accuracy of DNA
      replication, 12*
    exonuclease proofreading, 12
    mismatch repair, 13-14

replication and transcription
  factories, 38
subnuclear localization, 38
telomeres, 37
terminus regions, 37
DNA transfer, 45
DNA transposons, 41-42, 47-49
domains, 95-96, 153
double-strand breaks (DSBs),
  17, 153
Drosophila
  breakpoints, 114
  genome composition, 40
  genomic responses to changes in
    interspecific hybridization, 76
  Hox complex, 116
  transgenic fruit flies, 83
Ds transposon, 110
DSBs (double-strand breaks),
  17, 153
duplexes, 153
duplication of evolutionary
  inventions, 133-134
dysgenesis, 75

E

Escherichia coli
  accuracy of DNA replication, 12
  lambda prophage in, 60
  phase variation, 58
  sugar metabolism, 8-11
  Weigle mutagenesis, 15
ecological disruptions, 139-142
ectopic, 153
Ediacaran period, 117
EGF (epidermal growth factor),
  24, 153
endonuclease, 153
endoreplication, 154
endosymbiosis, 102-107, 154
Engels, Bill, 49
enhancers, 133, 154
ensuring transmission at cell
  division, 37
epidermal growth factor (EGF),
  24, 153
epigenetic imprinting sites, 111-112

epigenome, 33, 154
  definition, 80
  life history events that alter
    epigenome, 80-81
epigenetic regulation, 3, 31-36
erythromycin, 154
euchromatin, 34, 154
Euglenophyta, 103
eukaryotes, 133, 154
  antigenic variation, 56-59
  autonomous replication sequences
    (ars), 36
  centromeres, 37
  chloroplasts, 102
  DNA proofreading and repair
    systems, 14
  DNA transposons, 47-49
  evolution of, 100-107
  horizontal DNA transfer, 90-95
  HR (homologous recombination)
    explained, 45
    *role in phase and antigenic
      variation, 57*
  inteins, 54-55
  long terminal repeat (LTR)
    retroelements, 49-51
  mitochondria, 102
  phase variation, 56-59
  replication and transcription
    factories, 38
  retrosplicing group II introns, 53
  symbiogenesis, 100-107
  telomeres, 37
Evo-Devo, 116
Evolution by Gene Duplication
  (Ohno), 120
evolutionary advantages of targeting
  genome restructuring, 135-136
evolutionary breakpoints, 114
exchange, 154
excision repair, 16
exonization, 108-109
exons, 130-132, 154
exonuclease, 13, 154
export systems (DNA), 44-45
expression sites, 57, 154
extinctions, 139-140

F

factories, 38
FB (foldback) DNA transposons, 41
Fisher, R. A., 145
foldback (FB) DNA transposons, 41
formatting, 154
fossil record, correlation with Hox complex evolution, 116-120
fragile regions, 114
fragile sites, 114
Fugu rubripes, 40
fungi, RNA-based defense against viruses and plasmids, 78

G

G factor-coupled receptors, 133
G protein-coupled receptors, 22
gametes, 155
generation of novel DNA sequences, 131-133
"genes," inconsistent definition of, 29-30
Genetic Algorithm methods, 145
genetic imprinting, 32-33
genetic locus, 30
genome compaction, 31-36
genome components, 43
genome composition, 39
  classes of annotated repetitive genome components, 41-42
  classes of DNA in selected genomes, 40
genome regulation
  chromatin formatting, 31-36
  epigenetic regulation, 31-36
  transcriptional regulatory circuits, 30-31
genome restructuring. *See* natural genetic engineering
genome shock, 35, 69, 123
genomes, 155
genomic islands, 61, 94
genotoxicity, 18, 155
genotypes, 155
Geobacillus stearothermophilu, 58
geological boundaries of mass extinctions, 139-140

germ plasm, 155
germlines, 155
gonococci, 155
gradualism, 155
granules, 38
Greider, Carol, 64
Griffiths, Fred, 44
guide RNAs, 103
gypsy retrovirus, 51, 112

H

hairpin ends, 155
Haldane, J. B. S., 145
haploids, 155
Helicobacter pylori, 59
helitrons (rolling circle DNA transposons), 41, 155
heredity, 3-4
  epigenetic inheritance, 33
  genetic imprinting, 32
  paramutations, 33
heterochromatic silencing, 34
heterochromatin, 34, 155
heterocysts, 60
histone codes, 34
histones, 34-35, 156
Holland, John, 145
homeobox complex, 116-120
homeodomain proteins, 97
homeodomains, 108, 156
homing endonucleases, 54
Homo sapiens, 40
homologous recombination (HR), 17, 45-46, 57, 156
homologues, 156
homology, 156
Homopolymeric tract, 41
homopolymers, 156
Hooker, J. D., 122
horizontal DNA transfer, 90-95
host organisms, 156
Hox complex, 116-120, 130, 133, 156
HR (homologous recombination), 17, 45-46, 57, 156
hybrid dysgenesis, 75-78, 156
hybridization, 75-78, 121-122, 158
hybridogenesis, 105
hybrids, 156

hypersensitive response, 23, 157
hypha, 157

I

ICE (integrated conjugative
   element), 157
IES (internal eliminated sequences),
   63, 157
IGF (insulin-like growth factor),
   24, 157
imaginal disks, 118
immune system, 66-69
immunoglobulin class-switching, 136
import systems (DNA), 44-45
imprinting, 32-33, 157
indexing, 35
informatic view of living organisms,
   shift to, 4-5
inheritance
   epigenetic inheritance, 33
   genetic imprinting, 32
   paramutations, 33
initiating DNA replication 36
insulator bodies, 35
insulator sequences, 35, 157
insulin-like growth factor (IGF),
   24, 157
integrase, 50, 92, 157
integrated conjugative element
   (ICE), 157
integrons, 61, 92
inteins, 54-55, 157
intercellular signaling
   role in programmed celled death,
      23-24
   S. cerevisiae mating pheromone
      response, 21-22
interchromosomal exchange, 157
internal eliminated sequences (IES),
   63, 157
interphase, 158
interspecific hybridization, 75-78,
   121-122, 158
intracellular molecular information
   transfer, 24-26
introgression, 158
intronization, 109-110

introns, 132, 158
isotypes, 68, 158

J-K

Jacob, François, 132
junctional diversity, 67, 158

Karpechenko, Georgy, 121
kinetochore, 20, 158
kinetoplast mitochondrial
   genomes, 103
Kinetoplastida, 102
Kluyveromyces lactis, 62

L

lac operon (lactose operon), 9
   cooperative synergistic
      interactions, 131
   lacI, 11
   lacO, 10-11
   lacP, 10-11
lacI, 11
LacI repressor protein chain, 131
lacO, 10-11, 131
lacP, 10-11
Lactococcus lactis, 53
lactose operon (*lac* operon), 9-11,
   131
lambda prophage, site-specific
   recombination, 60
large-scale duplication (LSD),
   121, 159
larval transfer, 105
Legionella pneumophila, 58
Leishmania, 102
lichens, 100
life
   difficulty in defining, 128
   origin of, 128
ligation, 158
LINEs (long interspersed nucleotide
   elements), 42, 51-52, 65, 110, 158
linkage, 158
lipids, 159
localization, 38
locus, 159

long interspersed nucleotide
elements (LINEs), 42, 51-52, 65,
110, 158
long terminal repeat (LTR), 42,
49-51, 159
LSD (large-scale duplication),
121, 159
LTR (long terminal repeat), 42,
49-51, 159
Luria-Delbrück fluctuation test,
79-80
Lwoff, Andre, 49
lymphatic tissue, 159
lymphocytes, 29, 159
lymphokines, 68, 159
lysogeny, 159

M

macromolecules, 159
macronuclear destined sequences
(MDS), 63, 159
macronuclear development in ciliate
protozoa, 62-65
macronucleus, 63
MADS-box networks, 108
mammalian adaptive immune system,
66-69
mammalian nervous system, somatic
differentiation of, 65
MAPK kinase signaling cascades,
22, 133
Margulis, Lynn, 104
Marx, Groucho, 118
mass extinctions, 139-140
mating type switches in yeast, 61-62
maxicircles, 103
MB (mega base pairs), 159
McClintock, Barbara, 2, 35, 47, 110
MDS (macronuclear destined
sequences), 63, 159
mechanistic view of living organisms,
shift to informatic view, 4-5
MeCP2, 160
meiosis, 160
membranes, 160
metagenomics, 95
metazoa, 160

methanogenesis, 141
methyl groups, 160
methyl-directed mismatch repair
(MMR), 161
methylation, 160
microbe-generated atmospheric
transformations, 141
micronucleus, 63
microsatellites, 160
microtubules, 160
million years ago (MYA), 161
minicircles, 103
miRNA, 160
mismatch repair, 13-14
MITEs, 113
mitochondria, 102, 160
mitosis, 19, 160-161
mitotic spindle apparatus, 161
MMR (methyl-directed mismatch
repair), 161
model organisms, 101
Modern Evolutionary Synthesis, 142
modulators, 110
Monod, Jacques, 8-9
Moraxella bovis, 58
Moraxella lacunata, 58
morphogenesis, 161
multidrug resistance plasmids,
91, 161
Mus musculus, 40
mutagenesis, 14-16, 161
mutations, 1, 33
mutator phenotype, 13
mutator polymerase, 161
MutH protein, 13
MutL protein, 13
MutS protein, 13
mutualism, 161
MYA (million years ago), 161
Mycoplasma bovis, 59
Mycoplasma penetrans, 59
Mycoplasma pulmonis
antigenic variation, 59
phase variation, 58
Mycoplasma synoviae, 59
mycorrhiza, 101, 161

N

natural genetic engineering, 43-44
  antigenic variation, 56-59
  cellular regulation of, 69
    *cell operations that facilitate
    DNA restructuring, 43*
    *genomic responses to changes
    in ploidy and interspecific
    hybridization, 75-78*
    *interspecific hybridization,
    121-122*
    *life history events that alter the
    epigenome, 80-81*
    *RNA-based defense against
    viruses and plasmids, 78*
    *stimuli documented to activate
    natural genetic
    engineering, 70-74*
  chromosome rearrangements,
    113-115
  cis-regulatory modules (CRMs), 11,
    31, 110-111, 152
  definition of, 2, 161
  diversity-generating retroelements
    (DGRs), 55
    explained, 55
    *role in phase and antigenic
    variation, 57*
  DNA import and export systems,
    44-45
  DNA transposons, 47-49
  epigenetic imprinting sites and
    chromatin domains, 111-112
  exonization, 108-109
  evolutionary genomic
    innovation, 107
    *chromosome rearrangements,
    113-115*
    *cis-regulatory modules
    (CRMs), 110-111*
    *epigenetic imprinting sites and
    chromatin domains, 111-112*
    exonization, 108-109
    intronization, 109-110
    *regulating speed of
    transcripton, 110*
    *regulatory RNA molecules,
    112-113*
    *whole genome duplication
    (WGD), 114, 120-123, 126*

  *generation of novel DNA
    sequences, 131-133*
  genomes as read-write (RW)
    memory systems, 27-29
  genome regulation
    chromatin formatting, 31-36
    epigenetic regulation, 31-36
    *transcriptional regulatory
    circuits, 30-31*
  genome restructuring, 29
  homologous recombination
    (HR), 46
    explained, 45
    *role in phase and antigenic
    variation, 57*
  horizontal DNA transfer, 90-95
  Hox complex, 116-120
  hybridogenesis, 105
  implications of targeting genome
    restructuring, 134-139
  intronization, 109-110
  in lymphocytes, 29
  in normal organism life cycles, 55
    *macronuclear development in
    ciliate protozoa, 62-65*
    *mammalian adaptive immune
    system, 66-69*
    *mating type switches in yeast,
    61-62*
    *phase and antigenic variation,
    56-59*
    *somatic differentiation of
    mammalian nervous
    system, 65*
    *temperate bacteriophage life
    cycles, 60-61*
  inteins, 54-55
  links to ecological disruptions,
    139-142
  long terminal repeat (LTR)
    retroelements, 49-51
  morphological change by
    modification of genetic
    expression, 118-120
  non-homologous end-joining
    (NHEJ), 46
  non-LTR retroelements, 51-53
  origins in classical cytogenetics, 2
  phase variation, 56-59
  protein evolution, 95-98

rearrangements documented in
evolution of sequenced genomes,
123-124
regulating speed of
transcription, 110
regulatory RNA molecules, 112-113
reorganization of genome segments,
130-131
retention, duplication, and
diversification of evolutionary
inventions, 133-134
retrosplicing group II introns, 53-54
site-specific reciprocal
recombination
explained, 46-47
role in phase and antigenic
variation, 56-59
role in temperate
bacteriophage life cycles,
60-61
symbiogenesis and origin of
eukaryotes, 100-107
systems approach to generating
functional novelties, 129-130
targeting within genome
examples of, 84-86
explained, 82
P element homing, 83-84
21st century theory of
evolution, 142
basic evolutionary principles,
143-145
repercussions outside life
science, 145-147
whole genome duplication (WGD),
114, 120-123, 126
natural selection
chance and, 1-2
role in 21st century theory of
evolution, 144
ncRNA (noncoding RNA), 35,
112-113, 162
necrosis, 23
Neisseria gonorrhoea, 59
Neisseria meningitides, 58
A New Bacteriology (Sonea and
Panisset), 94
NHEJ (nonhomologous end-joining),
46, 162
nitrification, 141

nitrogen fixation, 141
non-LTR retroelements, 51-53
noncoding RNA (ncRNA), 35,
112-113, 162
nonhomologous end-joining (NHEJ),
46, 162
Nostoc, 55
nuclear pores, 162
nucleases, 162
nucleocytoplasmic large DNA
viruses, 92
nucleomorph, 103
nucleosomes, 34-35, 162
nucleotides, 162

O

Ohno, Susumu, 120
oligonucleotides, 162
oligonucleotide motif, 41
oligonucleotide recognition
sequences, 133
"one gene, one enzyme"
hypothesis, 28
ORC (origin of replication complex),
36, 162
ori sites, 36-37, 162
origin of life, 128
origin of replication (ori), 36-37, 162
origin of replication complex (ORC),
36, 162
origination, 162
Oryza sativa (indica), 40
Oryza sativa (Japonica), 40

P

P element homing, 83-84, 136
palindromes, 162
Panisset, Maurice, 94
par sites, 37, 163
paramutations, 33, 162
partition (par) sites, 37
penicillin, 163
peptide attachments, 163
peptide bonds, 163
peptide excision, 163
phages, 15, 150, 170
Phanerozoic era, 139
phase variation, 56-59, 163
phenotypes, 163

phosphodiester bonds, 163
phosphorylation, 11
photosynthesis, 141
phylogeny, 163
piRNA, 77, 163
piwi proteins, 163
plants
    genomic responses to changes in
        ploidy and interspecific
        hybridization, 75-78
    interspecific hybridization, 121-122
    plant neurobiology, 137
    RNA-based defense against viruses
        and plasmids, 78
plasmagenes, 101, 163
plasmids, 164
    multidrug resistance plasmids, 91
    plasmid replication, 60
    RNA-based defense against, 78
Plasmodium falciparum, 103
plastids, 102, 164
ploidy, 164
plus strand, 164
polymerization, 12, 164
polypeptide, 164
polyploid, 164
polytene chromosomes, 63, 164
position effect, 164
post-transcriptional, 164
POU domain circuits, 108
precision of DNA replication, 12
prions, 3
processed RNA, 164
programmed celled death,
    23-24, 164
prokaryotes, 133, 165
    antigenic variation, 56-59
    DNA transposons, 47-49
    horizontal DNA transfer, 90-95
    HR (homologous recombination)
        explained, 45-46
        *role in phase and antigenic
        variation, 57*
    inteins, 54-55
    origins of replication (ori sites), 36
    partition (par) sites, 37
    phase variation, 56-59
    retrosplicing group II introns, 53-54

site-specific reciprocal
    recombination
    explained, 46-47
    *role in phase and antigenic
    variation, 56-59*
    *role in temperate
    bacteriophage life cycles,
    60-61*
    terminus regions, 37
promoters, 165
proofreading DNA replication
    accuracy of DNA replication, 12
    exonuclease proofreading, 12
    mismatch repair, 13-14
prophage state, 60
prophages, 15, 165
Prosthecochloris, 55
proteins. *See also* specific proteins
    definition of, 165
    protein evolution, 95-98
    protein families, 97
    protein-DNA recognition, 10
proteolytic cleavage, 165
protists, 165
proviruses, 49, 60
Pseudoalteromonas atlanticus, 58
Pseudomonas fluorescens, 58
punctuated equilibrium, 144

Q-R

R&D (research and development)
    analogy, 132
RAD9 protein, 19, 165
radiation, 140
RAG proteins, 165
Raup, David, 139
rearrangement of chromosomes,
    113-115
RecA protein, 17, 165
reciprocal recombination,
    site-specific
    explained, 46-47
    role in phase and antigenic
    variation, 56-59
    role in temperate bacteriophage life
    cycles, 60-61
recombinase, 47, 165

regulation
  cellular regulation of natural genetic
    engineering, 69
    interspecific hybridization,
      121-122
    life history events that alter the
      epigenome, 80-81
    RNA-based defense against
      viruses and plasmids, 78
    stimuli documented to activate
      natural genetic engineering,
      70-74
  chromatin formatting, 31-36
  epigenetic regulation, 31-36
  regulatory RNA molecules, 112-113
  regulatory signals, 130
  transcriptional regulatory circuits,
    30-31
regulatory RNA molecules, 112-113
regulatory signals, 130
related species
  genetic similiarities between, 120
  morphological change by
    modification of genetic expression,
    118-120
reorganization of genome segments,
  130-131
repair (DNA), 14-16
  SOS response, 16-17
  Weigle mutagenesis, 15
repetitive DNA, 165
repetitive genome components,
  41-43
replication, 166
  centromeres, 37
  completion, 37
  ensuring transmission at cell
    division, 37
  initiation, 36
  origin of replication complex
    (ORC), 36
  origins of replication (ori sites), 36
  partition (par) sites, 37
  proofreading
    accuracy of DNA
      replication, 12
    exonuclease proofreading, 12
    mismatch repair, 13-14
  replication and transcription
    factories, 38

replication forks, 166
replication transposition, 166
subnuclear localization, 38
telomeres, 37
terminus regions, 37
replicons, 60, 166
research and development (R&D)
  analogy, 132
restructuring genomes. *See* natural
  genetic engineering
retention of evolutionary inventions,
  133-134
retroelements, 166
  long terminal repeat (LTR)
    retroelements, 49-51
  non-LTR retroelements, 51-53
retrohoming, 166
retrosplicing, 53-54, 166
retrotransduction, 52, 166
retrotransposition, 166
retrotransposons, 49, 166
retroviruses, 49-51, 167
reuse of evolutionary inventions,
  115-120
reverse transcriptase, 25, 167
ribosomes, 98, 167
ribozymes, 53, 167
RNA
  guide RNAs, 103
  reverse transcriptase activity, 25
  RNA-based defense against viruses
    and plasmids, 78
  RNA loops, 167
  RNA maturase, 167
  RNA Pol III, 167
rolling circle DNA transposons
  (helitrons), 41
Rous Sarcoma Virus (RSV)
  retroelement, 50
rRNA, 167
RSV (Rous Sarcoma Virus)
  retroelement, 50

S

S. cerevisiae mating pheromone
  response, 21-22
S. pneumoniae, 45
Saccharomyces cervisiae, 61-62, 167

Salmonella enterica, 58
Salmonella typhimurium, 18
saltationist, 167
SAM, 167
satellite DNA, 168
Schizosaccharomyces pombe,
    61-62, 168
second messengers, 10, 168
self-splicing, 168
Sepkowski, Jack, 139
serine recombinases, 47
7S RNA, 149
Shigella flexneri, 58
short interspersed nucleotide
    elements (SINEs), 42, 51-52,
    119-120, 168
shufflons, 57
signaling
    role in programmed cell death,
        23-24
    S. cerevisiae mating pheromone
        response, 21-22
    signal transduction, 168
silencers, 168
simple sequence repeats (SSR),
    41, 168
SINEs (short interspersed
    nucleotide elements), 42, 51-52,
    119-120, 168
siRNA, 168
site-specific reciprocal
    recombination
    definition of, 168
    explained, 46-47
    role in phase and antigenic
        variation, 56-59
    role in temperate bacteriophage life
        cycles, 60-61
small interfering RNA (siRNA), 168
small molecules, 168
somatic, definition of, 168
somatic differentiation of
    mammalian nervous system, 65
somatic hypermutation, 67, 136
Sonea, Sorin, 94
Sonneborn, Tracy, 3
SOS response, 16-17, 169
speed of transcription,
    regulating, 110
spindle checkpoints, 20-21

spindle poles, 169
splicing, 169
spores, 169
sprochetes, 169
SS (single-strand), 169
SSR (simple sequence repeats),
    41, 168
Staphylococcus aureus, 58
Staphylococcus epidermidis, 58
Stebbins, G. Ledyard, 121
Streptomycin, 169
stromatolites, 133
subnuclear localization, 38, 169
sugar metabolism in *E. coli*, 8-11
SulA protein, 18, 169
Sulfanilamide, 169
superintegrons, 61, 92
SVA elements, 52, 169
symbiogenesis, 100-107
syncytial, 169
syntenic, 113-114, 170
systems approach to generating
    functional novelties, 129-130
systems biology, 8, 170

T

T-DNA, 170
tandem array microsatellites, 41
tandem array satellites, 41
tandem repeat array, 170
target sites, 170
targeting of natural genetic
    engineering within genome, 83
    examples of, 84-86
    explained, 82
    P element homing, 83-84
taxonomy, 170
telomerase, 64, 170
telomeres, 37, 130, 133, 170
telophase, 170
Temin, Howard, 49
temperate bacteriophages,
    60-61, 170
template repeat (TR) segments, 55
templates, 171
terminal differentiation, 60, 171
terminal inverted repeat (TIR), 41,
    47, 171
terminal repeats, 171
terminal transferase, 171

: The Foxfire 45th anniversary book

:
ID: 0200903950455
: 1/31/2012

: The complete Hogan :
: McLean, Jim, 1950-
ID: 0200904090640
: 1/31/2012

: Evolution :
: Shapiro, James Alan, 1943-
ID: 0200904085707
: 1/31/2012

terminator sequences, 171
terminus regions, 37
tetracycline, 171
tetraploids, 171
thermophilic, 171
tinkering, 132
TIR (terminal inverted repeat), 41, 47, 171
tissue, 171
TNF (tumor necrosis factor), 24, 172
Toxoplasma gondii, 103
TR (template repeat) segments, 55
transcription
    coordinated transcription, 152
    definition of, 171
    regulating speed of, 110
    transcription factors, 133, 171
    transcription factories, 35, 172
    transcriptional regulatory circuits, 30-31
transfer of DNA, 45
transfer RNA (tRNA), 172
transforming principle, 44
transgenerational, 172
transgenic fruit flies, 83
translation, 172
transposase proteins, 48, 172
transposons, 41-42, 47-49, 172
Treponema pallidum, 55, 59
Trimethoprim, 172
Triticale interspecific hybrids, 121
tRNA, 172
21st century theory of evolution, 142
    basic evolutionary principles, 143-145
    repercussions outside life science, 145-147
Trypanosoma, 57, 102, 172
tumor necrosis factor (TNF), 24, 172
tyrosine recombinases, 47

U

undulipodia, 105, 172
uniformitarianism, 173
unique DNA sequences, 173
untemplated, 173
UV radiation, effect on mutagenic capacity in bacteria, 15

V

V(D)J joining, 173
v-src (viral oncogene), 50
variable nucleotide tandem repeats (VNTR), 41, 173
variable regions, 173
variable repeat (VR) region, 55
Vavilov, Nikolai, 121
vectors, 173
vesicles, 173
viral oncogene (v-src), 50
Virchow, Rudolph, 3
virosphere, 95
viruses, 173
    nucleocytoplasmic large DNA viruses, 92
    RNA-based defense against, 78
VNTR (variable nucleotide tandem repeats), 41, 173
VR (variable repeat) region, 55

W

Waddington, Conrad, 32
Weigle, Jean, 15, 146
Weigle mutagenesis, 15
Weismann, August, 155
Wessler, Susan, 110
whole genome duplication (WGD), 114, 120-123, 126, 173
Williamson, Donald, 105
Witkin, Evelyn, 16
Wnt, 174
Woese, Carl, 98
Wright, S., 145

X-Y-Z

Xanthomonas oryzae, 58

yeast
    mating pheromone response, 21-22
    mating type switches, 61-62
    RNA-based defense against viruses and plasmids, 78

Zea mays, 40
zygotes, 174

FT Press
FINANCIAL TIMES
SCIENCE

The life sciences revolution is transforming
our world as profoundly as the industrial
and information revolutions did
in the last two centuries.
FT Press Science will capture the excitement and
promise of the new life sciences, bringing breakthrough
knowledge to every professional and interested citizen.
We will publish tomorrow's indispensable work in
genetics, evolution, neuroscience, medicine,
biotech, environmental science, and whatever
new fields emerge next.
We hope to help you make sense of the future,
so you can *live* it, *profit* from it, and *lead* it.